Letts

GCSE

VISUAL REVISION GUIDE

SUCCESS

BIOLOGY

Author

Hannah Kingston

CONTENTS

HUMANS AS ORGANISMS

MAINTENANCE OF LIFE

INHERITANCE AND SELECTION

ENVIRONMENT

THE DIGESTIVE SYSTEM

- **This is really one long tube called the gut. If it were unravelled it would be about nine metres long!**
- **Digestion begins with the teeth and ends at the anus.**
- **It normally takes food 24–48 hours to pass through your digestive system.**

DIGESTION

- Digestion is the breaking down of <u>large</u>, <u>insoluble</u> molecules into <u>small</u>, <u>soluble</u> molecules so that they can be absorbed into the bloodstream.
- The large, insoluble molecules are starch, protein and fat.
- This action is speeded up (catalysed) by <u>enzymes</u>.
- Enzymes in the small intestine are found throughout the digestive system.

Large intestine receives any food that has not been absorbed into the blood. Excess water and salts are removed from the food. The remaining solid food is turned into <u>faeces</u>

Gall bladder is where <u>bile</u> is stored until it is released into the small intestine via the bile duct

Liver produces bile, a green solution that has two functions: – It <u>neutralises stomach acid</u> so that the enzymes in the small intestine can work properly. (Only pepsin likes acid conditions) – It acts on fats, breaking them up into small droplets. This is called <u>emulsification</u>. Emulsifying fats makes it easier for lipase enzymes to work as they have a larger surface area to work on

Note: Food does not pass through the <u>pancreas</u>, <u>liver</u> and <u>gall</u> <u>bladder</u>. They are organs that secrete enzymes and bile to help digestion

Mouth contains teeth that begin digestion by breaking up food

Salivary glands secrete amylase which is a <u>carbohydrase</u> <u>enzyme</u> – Mucus lubricates the food as it passes down the oesophagus

Oesophagus sometimes called the gullet

Stomach has muscular walls which churn up the food and mix it with the <u>gastric</u> <u>juices</u> that the stomach produces – The gastric juices contain protease enzymes and hydrochloric acid – The hydrochloric acid provides the acidic conditions for a protease enzyme called pepsin to work

Pancreas produces <u>carbohydrase</u>, <u>protease</u> and <u>lipase</u> <u>enzymes</u>

Small intestine also produces <u>all</u> <u>three</u> types of enzymes – This is where <u>digestion</u> is <u>completed</u> and dissolved food is <u>absorbed</u> into the bloodstream – The inner surface is covered in tiny finger like projections called <u>villi</u>

Rectum where the faeces are stored before they leave the body via the <u>anus</u>

HOW FOOD MOVES THROUGH THE GUT

- After swallowing, the food is pushed into your oesophagus (gullet).
- Your gut has circular muscles in its walls.

- These muscles contract and squeeze from behind the food to push it along.
- This contraction and relaxation of muscles is called <u>peristalsis</u>.

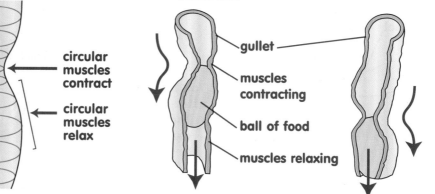

circular muscles contract

circular muscles relax

gullet

muscles contracting

ball of food

muscles relaxing

Examiner's Top Tip
Each part of the digestive system has a particular job. Learn the functions of each of the parts and where the enzymes and other helpful secretions are produced.

ABSORPTION AND THE VILLUS

- *The small intestine is where the digested food is absorbed into the blood.*
- *The small intestine is well designed for absorption.*
- *It has a <u>thin</u> <u>lining</u>, <u>a</u> <u>good</u> <u>blood</u> <u>supply</u> and <u>a</u> <u>very</u> <u>large</u> <u>surface</u> <u>area</u>.*
- *The large surface area is provided by the villi (single = villus) that extend from the inside of the small intestine wall.*

a villus is only one cell thick

amino acids, sugars, fatty acids and glycerol molecules are absorbed into the blood capillary

it contains a network of capillaries

blood arriving at the villus to pick up food molecules

blood leaving the villus, taking the food molecules to the rest of the body

Examiner's Top Tip
Make sure you can list the organs through which food passes on a complete journey through the digestive system.

QUICK TEST

1. Where is the enzyme amylase produced?
2. What are the functions of bile?
3. What is emulsification?
4. Where in the digestive system does the food get absorbed into the bloodstream?
5. What enzymes does the pancreas produce?
6. Where are the villi found?
7. By what process does food move through the digestive system?
8. Which organs produce enzymes?
9. Where is bile stored before it is released onto the food?
10. What are gastric juices?

10. Gastric juices are secreted into the stomach and contain a protease enzyme (called pepsin) and hydrochloric acid
9. Gall bladder
8. The salivary glands, pancreas, stomach and small intestine
7. Peristalsis
6. On the inside surface of the small intestine
5. Carbohydrases, proteases and lipases
4. Small intestine
3. The breaking down of fats into droplets to provide a larger surface area for the enzyme lipase to work
2. Neutralise stomach acid and break down fat into droplets
1. Salivary glands

TEETH BEGIN DIGESTION

There are four kinds of teeth, each have a role in <u>breaking up your food</u>:

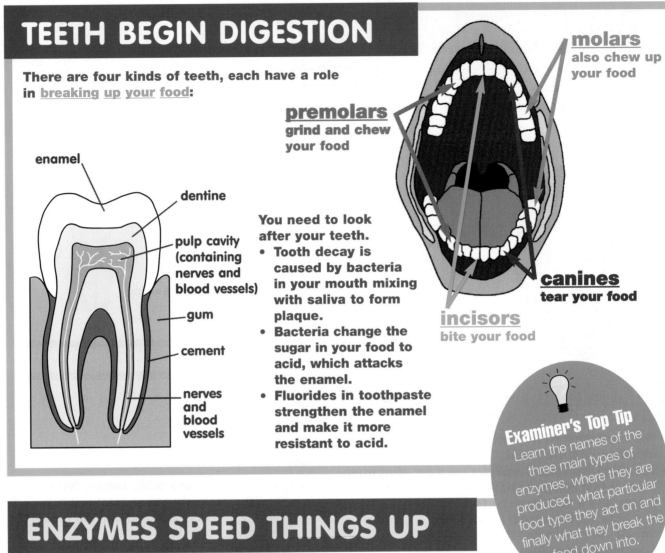

premolars
grind and chew your food

molars
also chew up your food

canines
tear your food

incisors
bite your food

enamel

dentine

pulp cavity (containing nerves and blood vessels)

gum

cement

nerves and blood vessels

You need to look after your teeth.
- Tooth decay is caused by bacteria in your mouth mixing with saliva to form plaque.
- Bacteria change the sugar in your food to acid, which attacks the enamel.
- Fluorides in toothpaste strengthen the enamel and make it more resistant to acid.

ENZYMES SPEED THINGS UP

- Starch, protein and fats are <u>large</u>, <u>insoluble</u> <u>food</u> molecules.
- Even after the teeth have done their bit and the stomach has churned the food up, it is still too big and insoluble to pass into the bloodstream.
- If you look back at the diagram of the digestive system you will see where these <u>chemicals</u> <u>called</u> <u>enzyme</u>s are made.
- Enzymes are specific. There are <u>three</u> <u>main</u> <u>enzymes</u> in your system.

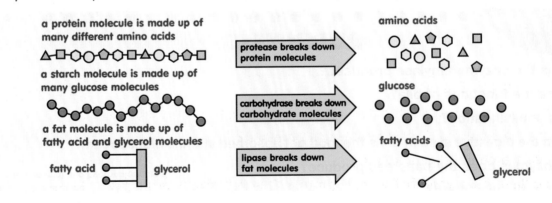

a protein molecule is made up of many different amino acids

a starch molecule is made up of many glucose molecules

a fat molecule is made up of fatty acid and glycerol molecules

fatty acid   glycerol

protease breaks down protein molecules

carbohydrase breaks down carbohydrate molecules

lipase breaks down fat molecules

amino acids

glucose

fatty acids

glycerol

- Starch is broken down into glucose in the mouth and small intestine.
- Proteins are broken down into amino acids in the stomach and the small intestine.
- Fats are broken down into fatty acids and glycerol in the small intestine.
- An example of a carbohydrase enzyme is amylase.
- An example of a protease enzyme is pepsin.

WHY DIGEST?

- Food has to be digested into <u>small</u> <u>and</u> <u>soluble</u> molecules, so that it can pass through the gut wall into the bloodstream.
- The blood carries the small, dissolved nutrients around the body to where its needed, the working cells.

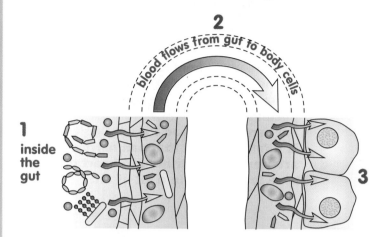

1. glucose, amino acids, fatty acids and glycerol are small enough to diffuse into the blood
2. the blood flows from the small intestine to the body cells
3. the food diffuses out again into the cells

EXPERIMENT

- You can prove that a carbohydrase called amylase breaks down starch into sugar by setting up the following experiment.

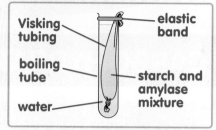

- After leaving the test-tube for 10 minutes in a water bath (maintained at 37°C), test the water for starch (add iodine), and for glucose (add Benedict's solution and heat).
- Visking tubing acts like a model gut; it has tiny holes in it that will only allow small molecules through.

Results
- The water should test negative for starch (yellow) and positive for sugar (orange).
- The results show that the starch has been broken down into glucose.
- The glucose is small enough to pass through the visking tubing into the surrounding water

HELP WITH DIGESTION

- **Ultimately we need nutrients from our food to keep our bodies healthy.**
- **Remember: Digestion breaks down large food molecules into small molecules so that they can pass into our blood stream.**
- **As food passes through the digestive system it needs help to break it down.**

QUICK TEST

1. Name the four types of teeth.
2. Name the type of enzyme that digests starch. Can you give an example?
3. Where are the enzymes that digest starch produced?
4. What does starch get digested into?
5. Name the type of enzyme that digests protein. Can you give an example?
6. Where are the enzymes that digest protein produced?
7. What does protein get digested into?
8. Name the enzyme that digests fats.
9. What do fats get broken down into?
10. What do large, insoluble food molecules need to be broken down into?

10. Small, soluble molecules
9. Fatty acids and glycerol
8. Lipase
7. Amino acids
6. Stomach, small intestine
5. Protease. Pepsin
4. Glucose
3. Mouth and small intestine
2. Carbohydrase. Amylase
1. Incisors, canines, premolars, molars

NUTRITION AND FOOD TESTS

CARBOHYDRATES

- Carbohydrates consist of starch and types of sugar e.g. glucose (the sugar our bodies use for respiration) and lactose (the sugar in milk).
- We need carbohydrates to <u>give us energy</u>.
- Starch is made up of smaller glucose molecules joined together. Plants store glucose as starch.
- Glycogen is also a carbohydrate. Animals store glucose as glycogen.

These foods contain a lot of carbohydrate:

CHEMICAL TEST FOR STARCH
- Add two drops of yellow <u>iodine solution</u> to food solution.
- Solution will turn blue/black if starch is present.

CHEMICAL TEST FOR GLUCOSE
- Add a few drops of <u>Benedict's solution</u> to food solution.
- Heat in a water bath until it boils.
- If glucose is present, an orange precipitate will form.

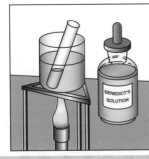

FATS

- Fats are made from fatty acids and glycerol.
- We need fats for a <u>store</u> of <u>energy</u>, to make <u>cell</u> <u>membranes</u> and for <u>warmth</u> (insulation).
- Fat can also be bad for us. <u>Cholesterol</u> is a fatty deposit that can narrow arteries and contribute to heart disease.

These foods contain a lot of fat:

CHEMICAL TEST FOR FAT
- Add 2 cm³ of <u>ethanol</u> to the food solution in a test tube and shake.
- Add 2 cm³ of <u>water</u> to the test-tube and shake again.
- Fat is present if the solution turns <u>cloudy white</u>.

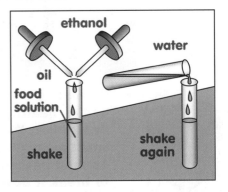

Examiner's Top Tip
Learn the food tests for starch, glucose, protein and fats.

WATER

Water makes up approximately 65% of your body weight.

Water is important because:
- Our blood plasma is mainly water.
- Water is in sweat that cools us down.
- Chemical reactions in our cells take place in water.
- Waste products are removed from our bodies in water.
- Food and drink contain water.

Examiner's Top Tip
Don't forget to learn examples of food belonging to each food group.

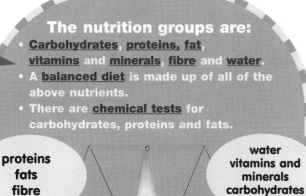

The nutrition groups are:
- <u>Carbohydrates</u>, <u>proteins</u>, <u>fat</u>, <u>vitamins</u> and <u>minerals</u>, <u>fibre</u> and <u>water</u>.
- A <u>balanced diet</u> is made up of all of the above nutrients.
- There are <u>chemical tests</u> for carbohydrates, proteins and fats.

proteins
fats
fibre

water
vitamins and
minerals
carbohydrates

A balanced diet

PROTEIN

- Your body cells are mostly made of protein.
- Proteins are made up of lots of amino acids.
- We need protein to <u>repair</u> and <u>replace</u> <u>damaged</u> cells or to make new cells during growth.

These foods containa lot of protein:

CHEMICAL TEST FOR PROTEIN, (THE BIURET TEST)

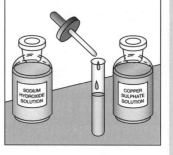

- Add some weak <u>copper sulphate</u> to the food solution.
- Carefully add drops of sodium hydroxide to the solution.
- If protein is present, the solution turns purple, gradually.

VITAMINS AND MINERALS

We only need these in small amounts, but they are essential for good health. Vitamins and minerals are found in fruit, vegetables and cereals. <u>Deficiency</u> <u>diseases</u> are caused by a lack of vitamins and minerals. <u>Vitamin C</u> keeps the skin strong and supple; without it the skin cracks and the gums bleed (called scurvy). <u>Vitamin D</u> helps the bones harden in children; without it the bones stay soft (a disease called rickets). We need the mineral <u>iron</u> for making haemoglobin and the mineral <u>calcium</u> for healthy bones and teeth.

FIBRE

- Fibre, or roughage, comes from plants.
- Fibre is not actually digested; it just keeps food moving smoothly through your system.
- Fibre provides something for your gut muscles to push against. It is a bit like squeezing toothpaste through a tube.
- It prevents constipation.

These are foods containing a lot of fibre:

Cereal

Examiner's Top Tip
Remember that carbohydrates are for energy, proteins are for growth and repair and fats are for energy store and insulation. But do not forget the other important food groups.

QUICK TEST

1. What do we use carbohydrates for?

2. Name the two main carbohydrates.

3. What is the chemical test for starch?

4. What is the chemical test for glucose?

5. What do our bodies need fat for?

6. How would you test for fat?

7. Why is protein important to our cells?

8. What is the chemical test for protein?

9. Why is fibre important?

10. What are deficiency diseases caused by?

1. Energy
2. Starch and glucose
3. Iodine; a blue/black colour means starch is present
4. Benedict's solution and heat; an orange precipitate means glucose is present
5. Store energy, make cell membranes and insulation
6. Ethanol and water; cloudy means fat is present
7. Repair and replace cells, and make new cells for growth
8. Biuret test; if solution turns purple, protein is present
9. It helps food move through your system and prevents constipation
10. Lack of vitamins and minerals

The blood follows a specific route through the heart and around the body. This is to ensure that all parts of the body get the substances they need and get waste substances removed. We have a **double circulation system**.
The blood passes through the heart twice on one circuit of the body. See Blood through the Heart.

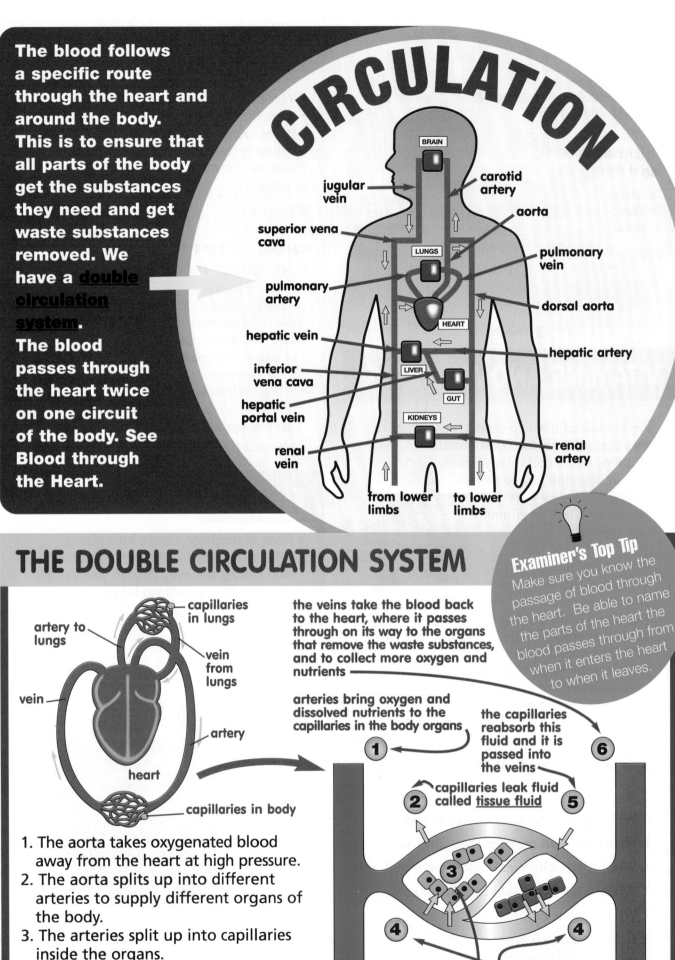

CIRCULATION

BRAIN
jugular vein
carotid artery
superior vena cava
aorta
LUNGS
pulmonary vein
pulmonary artery
dorsal aorta
hepatic vein
HEART
inferior vena cava
hepatic artery
hepatic portal vein
LIVER
GUT
KIDNEYS
renal vein
renal artery
from lower limbs
to lower limbs

Examiner's Top Tip
Make sure you know the passage of blood through the heart. Be able to name the parts of the heart the blood passes through from when it enters the heart to when it leaves.

THE DOUBLE CIRCULATION SYSTEM

capillaries in lungs
artery to lungs
vein from lungs
vein
artery
heart
capillaries in body

the veins take the blood back to the heart, where it passes through on its way to the organs that remove the waste substances, and to collect more oxygen and nutrients

arteries bring oxygen and dissolved nutrients to the capillaries in the body organs

1

the capillaries reabsorb this fluid and it is passed into the veins

6

2 capillaries leak fluid called **tissue fluid**

5

3

4   4

tissue fluid bathes the cells; this is where **exchange of substances** takes place

oxygen and **dissolved nutrients** diffuse into the cells and **carbon dioxide** and **water** diffuse out of the cells

1. The aorta takes oxygenated blood away from the heart at high pressure.
2. The aorta splits up into different arteries to supply different organs of the body.
3. The arteries split up into capillaries inside the organs.
4. The capillaries join up to form the veins leading out of the organs.
5. The veins join up to form the vena cava that transports deoxygenated blood back into the heart.

BLOOD THROUGH THE HEART

Remember the heart has two sides. The right side of your heart pumps deoxygenated blood from the body to the lungs and the left side pumps oxygenated blood from the lungs to the body.
There are three main stages to the heartbeat; it's a complete cycle so it can be started anywhere.

- Blood enters the two atria (singular = atrium) via either the pulmonary vein from the lungs or the vena cava from the body.
- The atria contract and push blood through the tricuspid and bicuspid valves into the ventricles.
- The ventricles contract, forcing blood into either the aorta (to go to the body) and the pulmonary artery (to go to the lungs). The semilunar valves snap shut to prevent backflow into the ventricles.

The oxygenated blood flows around the body in arteries and the deoxygenated blood returns in the veins to the atria, and the whole cycle begins again. One complete cycle is called a heartbeat. The left side of the heart has much thicker-walled ventricles as this side has to pump blood at high pressure all around the body.

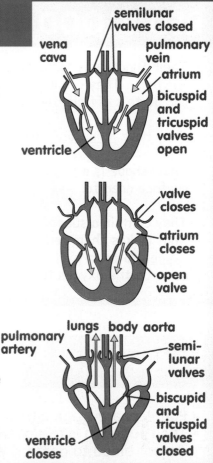

WHY DO WE NEED A TRANSPORT SYSTEM?

- Very small animals do not need a blood system; they can obtain the substances they need by simple diffusion.
- Larger animals need a transport system as the cells are far apart. Larger animals such as fish, insects and mammals also tend to be active and need rapid supplies of nutrients and oxygen.

HEART RATE AND EXERCISE

The heart would beat on its own outside of the body.
The pacemaker situated in the wall of the right atrium controls the heart rate. The pacemaker sends electrical messages to the heart muscle.
The pacemaker receives messages from the brain to either slow down heart rate or speed it up during exercise.
During exercise the heart pumps at a faster rate.
This ensures that the muscles receive oxygen as fast as possible and ensures the quick removal of carbon dioxide at the lungs. The blood is also diverted to where it is needed most, i.e. the muscles. Here the arteries get wider. The vessels supplying the less active organs, such as the gut, get narrower.

QUICK TEST

1. What is the name of the blood vessel entering the kidney?
2. Which side of the heart contains oxygenated blood?
3. After the ventricles, where does the blood go to?
4. Name the blood vessel that returns the blood to the heart.
5. Why is the heartbeat called a cycle?
6. Why does the blood go to the lungs?
7. Why is it called a double circulation?
8. What controls heart rate?
9. What blood vessels enter the organs and narrow down into capillaries?
10. Name the blood vessel to the head and arms.

Examiner's Top Tip
On the circulation diagram, trace the path of the blood through the heart to the organs and back again.

10. Carotid artery
9. Arteries
8. The pacemaker
7. Passes through the heart twice on one journey
6. To collect oxygen and release carbon dioxide
5. It is continuous
4. Vena cava
3. Out of the aorta and pulmonary artery
2. Left
1. Renal artery

RED BLOOD CELLS

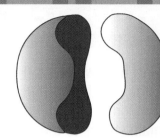

this diagram shows a red blood cell that has been sectioned to show its characteristic shape

- Their function is to carry <u>oxygen</u> to all the <u>cells of the body</u>. They are especially adapted so that they can do this efficiently.

- They have <u>no nucleus</u> (more room for oxygen).

- Their shape is a <u>biconcave disc</u>, which gives maximum surface area for <u>absorbing oxygen</u>.

- They contain a substance called <u>haemoglobin</u>. This is what makes them red. In the lungs it combines with <u>oxygen</u> to form <u>oxyhaemoglobin</u>. In the tissues it gives up the oxygen to form haemoglobin again.

Examiner's Top Tip
Learn the functions of the four parts of the blood and in particular how the structure of the red blood cell helps it do its job.

PLATELETS

- Platelets are <u>fragments of cells</u>.
- Their function is to clot the blood so you do not bleed to death if you cut yourself!

When you cut yourself, <u>platelets</u> form a <u>platelet plug</u> and along with a <u>protein</u> called <u>fibrinogen</u> (in the plasma) they form fibres, which seal up the blood leak. When it dries it forms a scab.

blood is made up of

white cells

red cells **platelets**

floating in a watery liquid called plasma

PLASMA

- Plasma is a yellow fluid.
- It consists of mainly water, but has many substances dissolved in it. These include <u>soluble food</u>, <u>salts</u>, <u>carbon dioxide</u>, <u>urea</u>, <u>hormones</u>, <u>antibodies</u> and <u>plasma proteins</u>.
- Its function is to transport these substances around the body.

FOR EXAMPLE:

<u>Soluble food</u> – from the small intestine to the liver for storage and all cells for respiration
<u>Urea</u> – from the liver to the kidneys for removal
<u>Carbon dioxide</u> – from the cells to the lungs for removal
<u>Plasma proteins</u> such as fibrinogen – travel to the site of a cut and along with the platelets clot the blood

- Plasma leaks out of the capillaries to form the <u>tissue fluid</u> that bathes the cells of the body. Tissue fluid contains all the substances that plasma does except proteins as these are too large to pass out of the capillaries.

WHITE BLOOD CELLS

- Their main function is <u>defence against disease</u>.

- They have a <u>large nucleus</u>.

- They are <u>larger</u> than red blood cells and their shape varies. There are two main types of white blood cells that can multiply if needed.

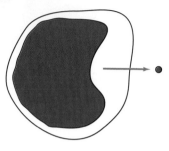

Phagocytes

this type of white blood cell kills germs by ingesting them

Lymphocytes

this type of white blood cell sends out antibodies which kill germs

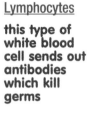

Phagocytes seek out and ingest (eat) microbes. They often die in the fight and the remains form 'pus' that bursts from a boil!

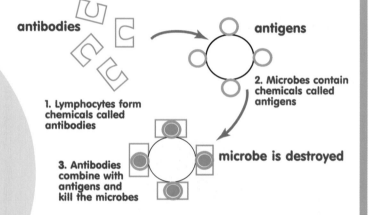

antibodies antigens

1. Lymphocytes form chemicals called antibodies

2. Microbes contain chemicals called antigens

3. Antibodies combine with antigens and kill the microbes

microbe is destroyed

Not all microbes are killed instantly, which is why you sometimes feel ill. The white blood cells multiply and gradually overcome them. The white blood cells also have a memory so that if the same microbe enters the body they can destroy them before it does any damage. This is called <u>immunity</u>.

BLOOD

THE RIVER OF LIFE CONSISTS OF <u>RED BLOOD CELLS</u>, <u>WHITE BLOOD CELLS</u> AND <u>PLATELETS</u> SUSPENDED IN A FLUID CALLED <u>PLASMA</u>

QUICK TEST

1. What are the four main components of blood?

2. Name two types of white blood cell.

3. What is the function of the white blood cells?

4. What do platelets do and what substance helps them?

5. Which types of cell have no nucleus?

6. Name the substance in red blood cells that combines with oxygen.

7. What is the substance in question five called when it combines with oxygen?

8. Name five substances that are dissolved in plasma.

9. Which type of white blood cell ingest bacteria?

10. Which type of white blood cell produce antibodies?

10. Lymphocytes
9. Phagocytes.
8. Urea, hormones, soluble food, carbon dioxide, plasma proteins, salts, hormones (any five)
7. Oxyhaemoglobin
6. Haemoglobin
5. Red blood cells
4. Clot blood; fibrinogen
3. Fight disease
2. Phagocytes, lymphocytes
1. Plasma, red blood cells, white blood cells, platelets

VEINS

vein

thin wall

lumen

- Veins carry <u>deoxygenated</u> blood.
- The <u>pulmonary</u> <u>vein</u> is the only vein to <u>carry</u> <u>oxygenated</u> <u>blood</u>. This is because it has just been to the lungs. Find it on the diagram.
- They carry the blood <u>back</u> <u>to</u> <u>the</u> <u>heart</u> from the body at low pressure.
- They have <u>valves</u> to prevent the blood flowing backwards.

longitudinal section of vein

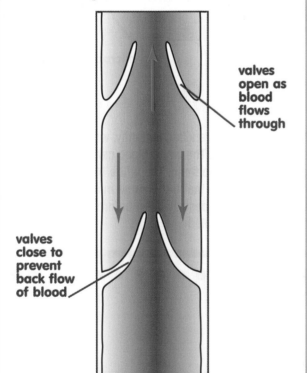

valves open as blood flows through

valves close to prevent back flow of blood

THE HEART

- The heart is made of a special type of muscle called cardiac muscle. It contracts continuously without getting tired.
- It is a double pump.
- The right side pumps blood to the lungs to be <u>oxygenated</u>.
- The left side pumps blood around the body and it becomes <u>deoxygenated</u> as it drops off oxygen to the tissues.

VEINS = IN

CAPILLARIES

- **Capillaries are only <u>one</u> <u>cell</u> <u>thick</u> and have very thin walls, to allow oxygen and nutrients to diffuse out of them.**
- **They are the site of exchange between the blood and the cells of the body.**
- **Fluid leaks out of the capillaries and bathes the surrounding cells. This is called <u>tissue</u> <u>fluid</u>.**
- **The tissue fluid delivers <u>nutrients</u> and <u>oxygen</u> to the cells and takes away waste products.**

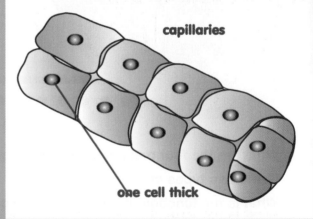

capillaries

one cell thick

CORONARY HEART DISEASE

- The coronary arteries supply the heart with oxygen and nutrients.
- Excess cholesterol, alcohol, stress and smoking all contribute to blocking these arteries.
- Excess cholesterol can 'fur' up the arteries and block blood flow. This can result in a heart attack.

ARTERIES

- Arteries carry <u>oxygenated</u> blood.
- The <u>pulmonary artery</u> is the only artery to carry <u>deoxygenated blood</u>. This is because it is going to the lungs to pick up oxygen. Find it on the diagram.
- They carry blood <u>away</u> from the heart towards the body at <u>high</u> pressure.
- They have very <u>thick</u>, <u>elastic</u> <u>walls</u> to withstand the high pressure.
- The high pressure in the arteries causes a <u>pulse</u> that can be felt especially in the wrist and neck.
- Arteries narrow down into capillaries.

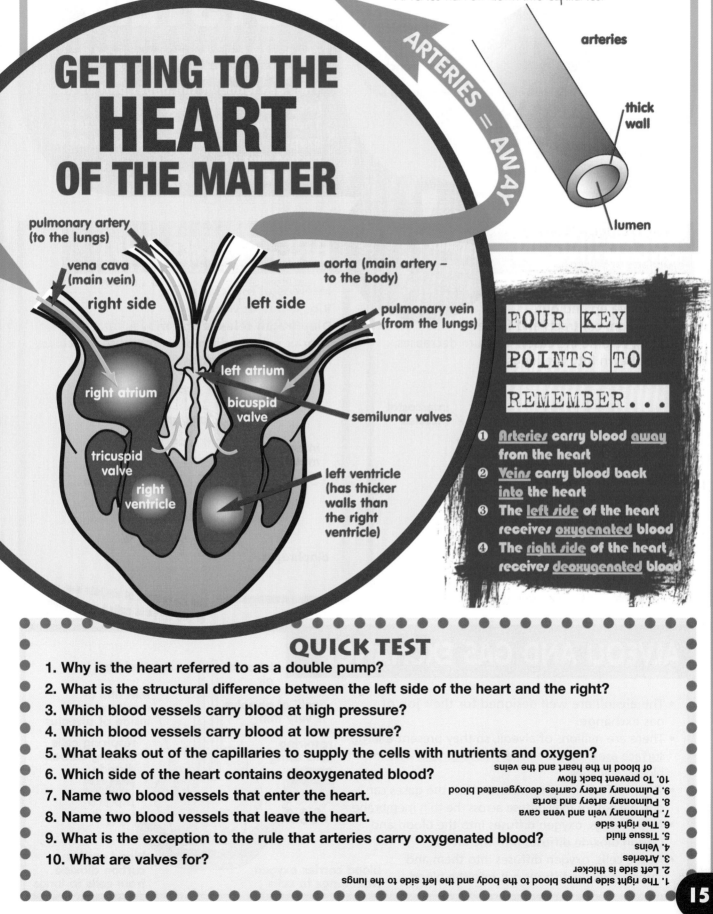

GETTING TO THE HEART OF THE MATTER

ARTERIES = AWAY

arteries

thick wall

lumen

pulmonary artery (to the lungs)

vena cava (main vein)

right side

aorta (main artery – to the body)

left side

pulmonary vein (from the lungs)

left atrium

bicuspid valve

right atrium

semilunar valves

tricuspid valve

right ventricle

left ventricle (has thicker walls than the right ventricle)

FOUR KEY POINTS TO REMEMBER...

① <u>Arteries</u> carry blood <u>away</u> from the heart
② <u>Veins</u> carry blood back <u>into</u> the heart
③ The <u>left side</u> of the heart receives <u>oxygenated</u> blood
④ The <u>right side</u> of the heart receives <u>deoxygenated</u> blood

QUICK TEST

1. Why is the heart referred to as a double pump?
2. What is the structural difference between the left side of the heart and the right?
3. Which blood vessels carry blood at high pressure?
4. Which blood vessels carry blood at low pressure?
5. What leaks out of the capillaries to supply the cells with nutrients and oxygen?
6. Which side of the heart contains deoxygenated blood?
7. Name two blood vessels that enter the heart.
8. Name two blood vessels that leave the heart.
9. What is the exception to the rule that arteries carry oxygenated blood?
10. What are valves for?

1. The right side pumps blood to the body and the left side to the lungs
2. Left side is thicker
3. Arteries
4. Veins
5. Tissue fluid
6. The right side
7. Pulmonary vein and vena cava
8. Pulmonary artery and aorta
9. Pulmonary artery carries deoxygenated blood
10. To prevent back flow of blood in the heart and the veins

THE BREATHING SYSTEM

The breathing system consists of the <u>lungs</u> and <u>diaphragm</u>.
They are found in the upper part of the body called the <u>thorax</u>.
Their function is to breathe in air to get <u>oxygen</u> and breathe out air to get rid of <u>carbon dioxide</u>.
This is called <u>gas exchange</u>.

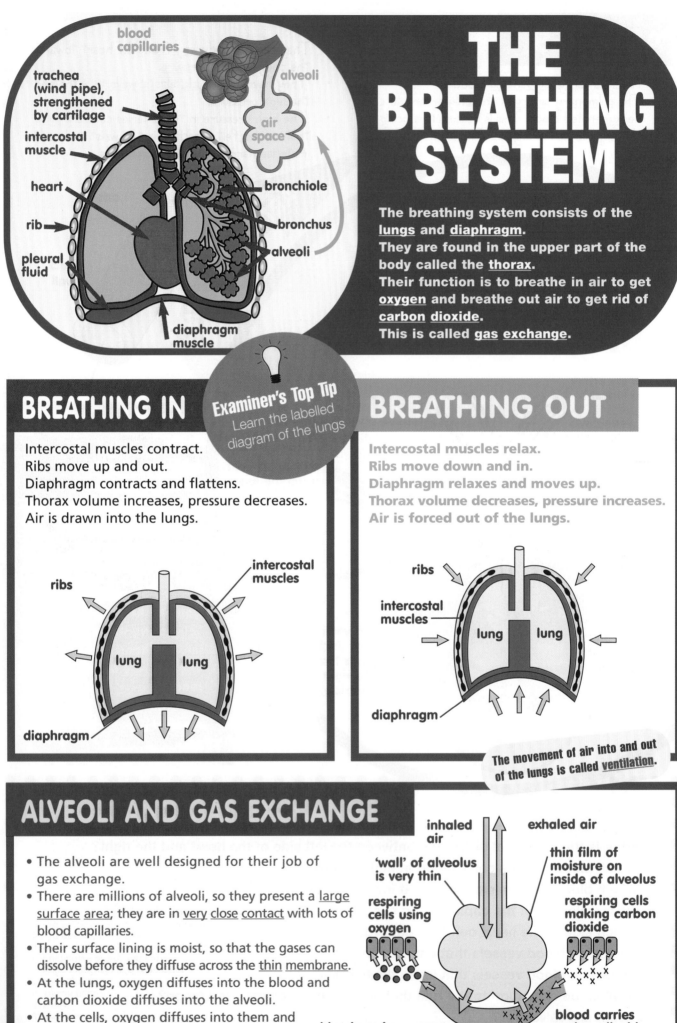

blood capillaries

trachea (wind pipe), strengthened by cartilage

alveoli

air space

intercostal muscle

heart

bronchiole

bronchus

rib

alveoli

pleural fluid

diaphragm muscle

Examiner's Top Tip
Learn the labelled diagram of the lungs

BREATHING IN

Intercostal muscles contract.
Ribs move up and out.
Diaphragm contracts and flattens.
Thorax volume increases, pressure decreases.
Air is drawn into the lungs.

ribs

intercostal muscles

lung lung

diaphragm

BREATHING OUT

Intercostal muscles relax.
Ribs move down and in.
Diaphragm relaxes and moves up.
Thorax volume decreases, pressure increases.
Air is forced out of the lungs.

ribs

intercostal muscles

lung lung

diaphragm

The movement of air into and out of the lungs is called <u>ventilation</u>.

ALVEOLI AND GAS EXCHANGE

• The alveoli are well designed for their job of gas exchange.
• There are millions of alveoli, so they present a <u>large</u> <u>surface</u> <u>area</u>; they are in <u>very</u> <u>close</u> <u>contact</u> with lots of blood capillaries.
• Their surface lining is moist, so that the gases can dissolve before they diffuse across the <u>thin</u> <u>membrane</u>.
• At the lungs, oxygen diffuses into the blood and carbon dioxide diffuses into the alveoli.
• At the cells, oxygen diffuses into them and carbon dioxide diffuses out into the blood.

inhaled air

exhaled air

'wall' of alveolus is very thin

thin film of moisture on inside of alveolus

respiring cells using oxygen

respiring cells making carbon dioxide

blood carries oxygen from lungs to cells

blood carries carbon dioxide from cells to lungs

COMPOSITION OF GASES

IN INHALED AIR	IN EXHALED AIR
Oxygen – 21%	Oxygen – 16%
Carbon dioxide – 0.04%	Carbon dioxide – 4%
Nitrogen – 79%	Nitrogen – 79%
Water vapour – a little	Water vapour – a lot

NOTE
- Notice that we breathe out oxygen and carbon dioxide as well as breathing them both in.
- It is important to note that we breathe in <u>more</u> oxygen and breathe out <u>more</u> carbon dioxide.
- You should also notice that the amount of water vapour is higher in the air that we breathe out. There are other important differences.

WHY BREATHE?

Breathing is necessary for respiration.
Cells need oxygen to produce energy from glucose.
Cells produce carbon dioxide as a waste product.

SMOKING AND LUNG DISEASE

- Tar contained in tobacco smoke can cause cancer of the lung cells. It can also irritate air passages and make them narrower, causing a 'smoker's cough'.
- Bronchitis is aggravated by smoking. Smoke irritates the air passages making them inflamed. The cilia stop beating, so mucus collects in the lungs along with dirt and bacteria.
- Emphysema is when the chemicals in tobacco smoke weaken the alveoli walls. The lung tissue can become damaged and make breathing difficult.

bronchiole
alveolus (air sac)

surface area greatly reduced

healthy person

person suffering from emphysema

CILIATED EPITHELIAL CELLS

- *We warm, moisten, filter and clean the air we breathe in.*
- *Lining the nose and lungs are cells that make a slimy liquid called <u>mucus</u>.*
- *Dust and germs get trapped in the mucus and tiny hairs called <u>cilia</u> waft up to the nose and throat to prevent them reaching the lungs.*
- *The air we breathe out is actually cleaner.*

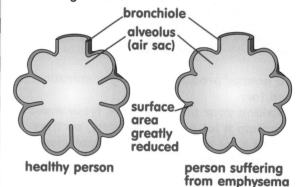

ciliated epithelial cells

hairs move mucus along

these cells make mucus

QUICK TEST

1. Where does gas exchange take place?
2. What happens to the ribs when you breathe out?
3. What happens to the diaphragm when you breathe in?
4. Why are the alveoli so good at gas exchange?
5. Why is breathing important?
6. What do the bronchi branch up into?
7. What strengthens the trachea?
8. What do the ciliated cells produce that traps dirt and bacteria?
9. What are cilia?
10. Give three differences between the air we breathe out and the air we breathe in.

1. The alveoli
2. Move down and out
3. It contracts and flattens
4. Large surface area, moist, thin walls, close to blood capillaries
5. For respiration
6. Bronchioles
7. C-shaped cartilage
8. Mucus
9. Tiny hairs on mucus making cells
10. Air out is cleaner, warmer, contains more water vapour and carbon dioxide (any three)

AEROBIC RESPIRATION

Aerobic means 'with air' and as respiration needs oxygen, we call it **aerobic respiration**. The chemical equation for respiration is:

- $C_6H_{12}O_6 + 6O_2 \Rightarrow 6CO_2 + 6H_2O + $ **ENERGY**

And the word equation is:

- Glucose + Oxygen ➡ Carbon dioxide + Water + **ENERGY**

The **carbon dioxide** and **water** are **waste products**, removed from the body in via the lungs, skin and kidneys.

USES OF THE ENERGY PRODUCED

The energy produced during respiration is used for:

- **Making your muscles work (your muscles contain a lot of mitochondria, as they need a lot of energy during exercise).**
- **Absorbing molecules against concentration gradients – called active transport.**
- **Chemical reactions.**
- **Growth and repair of cells.**
- **Making up larger molecules from smaller ones, i.e. proteins from amino acids.**
- **Maintaining body temperature in warm-blooded animals.**

SIMILARITIES BETWEEN RESPIRATION AND BURNING

Respiration inside every living cell is similar to burning – except there are no flames. They both require a fuel and oxygen, and they both release energy and waste products.

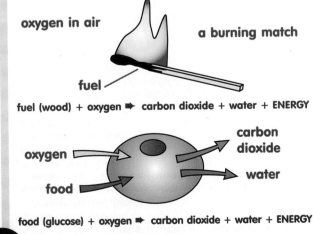

oxygen in air   a burning match

fuel

fuel (wood) + oxygen ➡ carbon dioxide + water + ENERGY

oxygen   carbon dioxide

food   water

food (glucose) + oxygen ➡ carbon dioxide + water + ENERGY

ANAEROBIC RESPIRATION

Respiration <u>without</u> <u>oxygen</u> is called <u>anaerobic respiration</u>. It produces <u>much less energy</u> and does not break down glucose completely.
This is the word equation:

- Glucose ➡ Energy + Lactic acid

Anaerobic respiration occurs when there is not enough oxygen available but the body still needs energy to move.
This might be during vigorous exercise.
Instead of carbon dioxide, <u>lactic acid</u> is produced.
Lactic acid builds up in muscles and causes them to ache and develop muscle cramps.
Fast, deep breathing as you recover soon supplies the body with enough oxygen to combine with the lactic acid to make carbon dioxide and water.
The amount of oxygen needed to remove the lactic acid is called the <u>oxygen debt</u>.

Examiner's Top Tip
The word equations and the chemical equations of respiration are both important to learn.

RESPIRATION AND PHOTOSYNTHESIS

The equations are the same but in opposite directions.

<u>Respiration</u>:
Glucose + Oxygen ➡ Carbon dioxide + Water

<u>Photosynthesis</u>:
Carbon dioxide + Water ➡ Glucose + Oxygen

- Plants carry out both reactions: photosynthesis using energy from the Sun to make glucose and oxygen in the day; and respiration in both the day and night.
- Animals use glucose and oxygen which the plants produce to carry out respiration.
- There is less carbon dioxide produced in the day as the plants use it up, but at night there is a rise in carbon dioxide as both plants and animals are respiring.
- Overall, plants use up more carbon dioxide than they produce, which stops the level of carbon dioxide increasing in the atmosphere.

RESPIRATION

Respiration is not breathing in and out.
Respiration is the <u>breakdown</u> of <u>glucose</u> to make <u>energy using oxygen</u>.
Every living cell in every living organism uses respiration to make energy, all of the time.
Energy is needed for all the chemical reactions in the body.
Respiration takes place in the <u>mitochondria in</u> the <u>cell cytoplasm</u>.

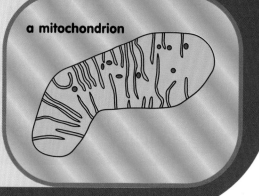

a mitochondrion

ANAEROBIC RESPIRATION IN YEAST

- *In yeast anaerobic respiration is called <u>fermentation</u>.*
- *It is summarised by the following equation:* glucose ➡ alcohol + carbon dioxide + <u>energy</u>
- *Uses of fermentation include making bread rise, making alcoholic drinks such as beer, wine and cider.*

MEASURING THE ENERGY CONTENT IN FOOD

- We can measure how much energy is in food by burning a sample of it and seeing how much energy it releases in heating up a volume of water.
- The temperature of the water is taken at the beginning.
- The food substance, in this case a peanut, is set alight and held underneath the water.
- When the peanut has stopped burning the temperature of the water is taken again. The difference in temperature is recorded.
- You can compare different food substances to see how much energy is inside them using this method, though there are lots of ways to improve it.

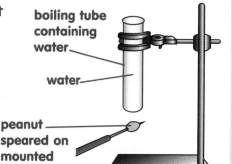

boiling tube containing water

water

peanut speared on mounted needle

QUICK TEST

1. Define aerobic respiration.
2. Define anaerobic respiration.
3. What is anaerobic respiration in yeast called?
4. How is lactic acid produced?
5. How is lactic acid removed?
6. Where does respiration take place?
7. What are the waste products of respiration?
8. Where does the oxygen come from that enables animals to carry out respiration?
9. When do plants carry out respiration?
10. Where do plants get the glucose and oxygen they need for respiration?

Examiner's Top Tip
How lactic acid is produced and how it is removed are using the oxygen debt are typical exam questions.

10. From photosynthesis, using energy from the Sun
9. All the time
8. From plants photosynthesising
7. Water and carbon dioxide
6. Cytoplasm of cells, in the mitochondria
5. By replacing the oxygen debt, fast and deep breathing
4. During vigorous exercise, without enough oxygen
3. Fermentation
2. Incomplete breakdown of glucose without oxygen
1. Breaking down glucose with oxygen to make energy.

THE NERVOUS SYSTEM

The nervous system is in charge. It <u>controls</u> and <u>co-ordinates</u> the parts of your body so that they work together at the right time.
The nervous system co-ordinates things you don't even think about like breathing and blinking.

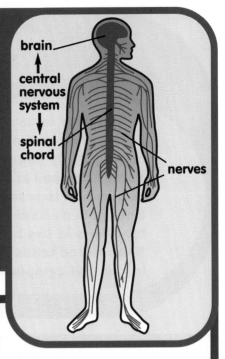

brain
central nervous system
spinal chord
nerves

THE CENTRAL NERVOUS SYSTEM

- <u>The central nervous system</u> (CNS) consists of the brain and spinal cord connected to different parts of the body by <u>nerves</u>.
- Your body's sense organs contain <u>receptors</u>.
- Receptors detect changes in the environment called stimuli.

<u>Nose</u>	~ sensitive to chemicals in the air
<u>Mouth</u>	~ sensitive to chemicals in food
<u>Ears</u>	~ sensitive to sound and balance
<u>Skin</u>	~ sensitive to touch, pressure and temperature
<u>Eyes</u>	~ sensitive to light

- The receptors send messages along nerves to the brain and spinal cord in response to stimuli from the environment.
- The messages are called nerve impulses.
- The CNS sends <u>nerve impulses</u> back along nerves to <u>effectors</u> which bring about a response.
- Effectors are muscles that bring about movement, or glands that secrete hormones.

NERVES

Nerves are made up of nerve cells or <u>neurones</u>.
There are three types of neurone.
Neurones have a nucleus, cytoplasm and cell membrane, but they have changed their shape and become specialised.

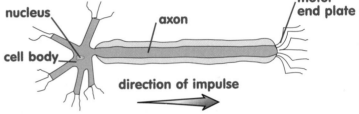

nucleus
axon
motor end plate
cell body
direction of impulse

The <u>sensory neurones</u> receive messages from the receptors and send them to the CNS.

nucleus
dendron
cell body
direction of impulse
fatty sheath
axon

The <u>motor neurones</u> send messages from the CNS to the effectors telling them what to do.
Nerve impulses travel in <u>one direction only</u>.
The fatty sheath is for insulation and for speeding up nerve impulses.
A <u>relay neurone</u> connects the sensory neurone to the motor neurone in the CNS.

DISEASES OF THE NERVOUS SYSTEM

- <u>Multiple sclerosis</u> is caused by the fatty sheath around the neurone breaking down. Without the fatty sheath impulses slow down or even stop, preventing the messages reaching the muscles.
- <u>Motor neurone disease</u> affects the neurones that join the muscles.
The neurones break down and cannot transmit nerve impulses. The muscles cannot contract and the person becomes paralysed.

SYNAPSES

- In between the neurones there is a gap called a synapse.
- When an impulse reaches the end of an axon a chemical is released.
- This chemical diffuses across the gap.
- This starts off an impulse in the next neurone.
- Synapses can be affected by drugs and alcohol, which slow down synapses or even stop them.

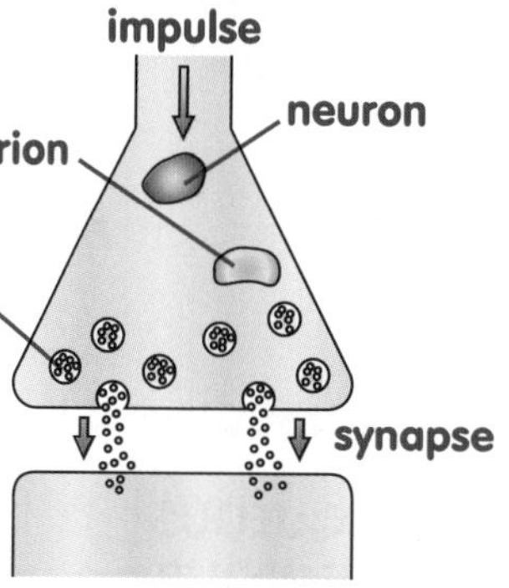

impulse
neuron
mitochondrion
chemical transmitter
synapse

chemical diffuses across synapse to start impulse in next neuron

THE REFLEX ARC

white matter (nerve fibres) ganglion
spinal chord synapse
nerve impulses from pain
pain receptor
sensory nerve cell body
sensory nerve fibre
motor nerve fibre
grey matter (neurone cell bodies)
synapse
relay nerve fibre
muscle
motor end-plate (motor nerve ending inside muscle)

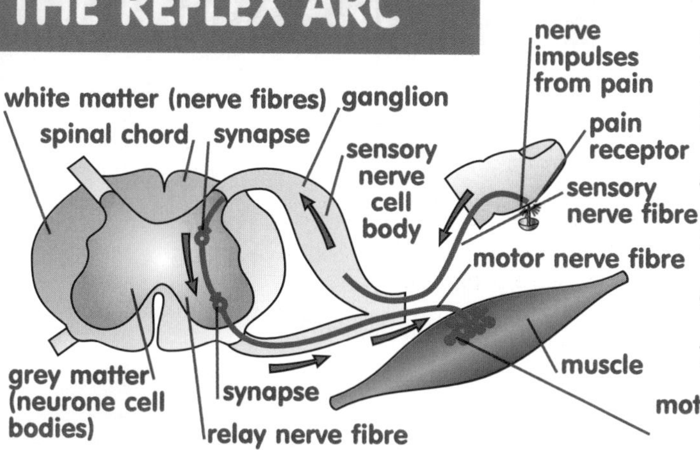

The reflex response to your CNS and back again can be shown in a diagram called the reflex arc.
1 Stimulus in this example is a sharp object.
2 The receptor is the pain sensor in the skin.
3 The nerve impulse travels along the sensory neurone.
4 The impulse is passed across a synapse to the relay neurone.
5 The impulse is passed across a synapse to the motor neurone.
6 The impulse is passed along a synapse to the muscle effector in the arm.
7 You move your hand away.

The reflex arc can be shown in a block diagram:

Stimulus ⇒ Receptor ⇒ Sensory neurone ⇒ Relay neurone ⇒ Motor neurone ⇒ Effector ⇒ Response

REFLEX AND VOLUNTARY ACTIONS

- Voluntary actions are things you have to think about, like talking or writing. They have to be learned.
- Reflex actions happen so quickly you haven't got time to think, they often protect you. They are automatic: examples include blinking and the knee-jerk reflex.

Examiner's Top Tip
Make sure you learn the diagram of the reflex arc and the block diagram, either one may come up in the exam.

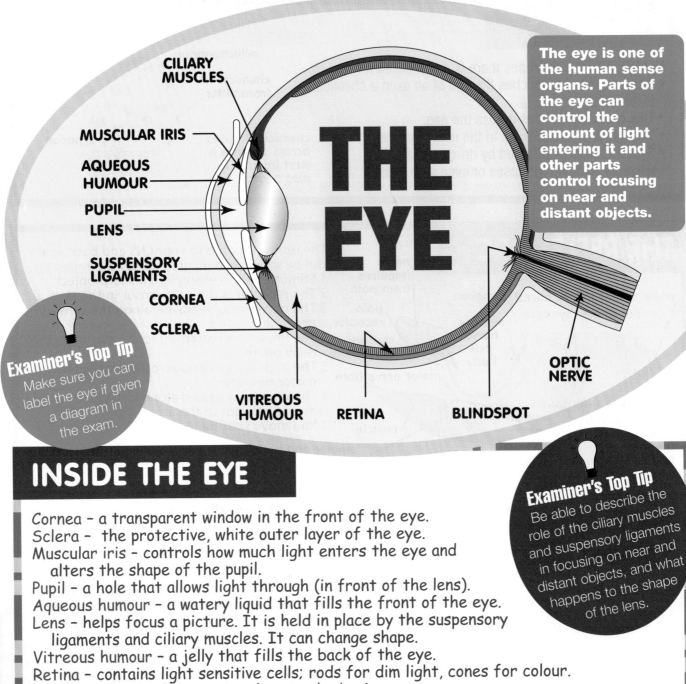

THE EYE

CILIARY MUSCLES

MUSCULAR IRIS

AQUEOUS HUMOUR

PUPIL

LENS

SUSPENSORY LIGAMENTS

CORNEA

SCLERA

VITREOUS HUMOUR

RETINA

BLINDSPOT

OPTIC NERVE

The eye is one of the human sense organs. Parts of the eye can control the amount of light entering it and other parts control focusing on near and distant objects.

Examiner's Top Tip
Make sure you can label the eye if given a diagram in the exam.

Examiner's Top Tip
Be able to describe the role of the ciliary muscles and suspensory ligaments in focusing on near and distant objects, and what happens to the shape of the lens.

INSIDE THE EYE

Cornea – a transparent window in the front of the eye.
Sclera – the protective, white outer layer of the eye.
Muscular iris – controls how much light enters the eye and alters the shape of the pupil.
Pupil – a hole that allows light through (in front of the lens).
Aqueous humour – a watery liquid that fills the front of the eye.
Lens – helps focus a picture. It is held in place by the suspensory ligaments and ciliary muscles. It can change shape.
Vitreous humour – a jelly that fills the back of the eye.
Retina – contains light sensitive cells; rods for dim light, cones for colour. The retina sends nerve impulses to the brain.
Blind spot – where blood vessels and nerves join the eyeball.
Optic nerve – receives nerve impulses from the retina and sends them to the brain.
Ciliary muscles – change the thickness of the lens when focusing.
Suspensory ligaments – hold the lens in place.

SEEING THINGS

Light from an object enters the eye through the cornea.
The curved cornea and lens produce an image on the retina that is upside down.
The receptor cells in the retina send impulses to the brain along sensory neurones in the optic nerve.
The brain interprets the image and you see the object the right way up.

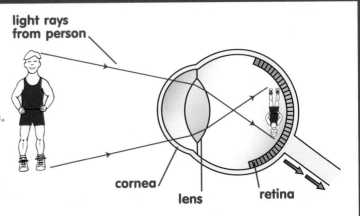

light rays from person

cornea

lens

retina

ADJUSTING TO LIGHT AND DARK

BRIGHT LIGHT
Circular muscles contract. Radial muscles relax.
The iris closes and makes the pupil smaller.
Less light enters the eye.

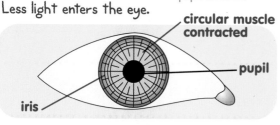

circular muscle
contracted

pupil

iris

DIM LIGHT
Radial muscles contract. Circular muscles relax.
The iris opens and makes the pupil bigger.
More light enters the eye.

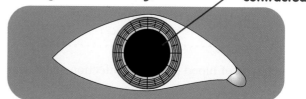

radial muscle
contracted

FOCUSING ON DISTANT OBJECTS

- **The ciliary muscles <u>relax</u>.**
- **This causes the suspensory ligaments to pull <u>tight</u>.**
- **The lens gets pulled <u>thin</u> and <u>flat</u> and only bends light a little.**
- **The distant object is focused on the retina.**

light from distant object

thin lens

light focused on retina

ciliary muscles relaxed

FOCUSING ON NEAR OBJECTS

The ciliary muscles <u>contract</u>.
This causes the suspensory ligaments to <u>slacken</u>.
The lens gets <u>fatter</u> and <u>rounder</u>, which bends light a lot.
The near object is focused on the retina.

light from near object

thicker lens

light focuses on retina

ciliary muscles contracted

QUICK TEST

1. Name the part of the eye that controls the amount of light entering it.

2. What is the name of the hole in the middle of the iris?

3. Which part of the eye contains the light sensitive cells?

4. What happens to the size of the pupil in bright light?

5. What happens to the size of the pupil in dim light?

6. Name the muscles that control the size of the lens?

7. What is the function of the optic nerve?

8. What shape is the lens when focusing on near objects?

9. What shape is the lens when focusing on distant objects?

10. Which part of the eye does light enter through?

10. The cornea
9. Thin and flat
8. Fat and round
7. Receives nerve impulses from the retina and sends them to the brain
6. Ciliary muscles
5. It gets larger
4. It gets smaller
3. Retina
2. Pupil
1. Iris

EXAM QUESTIONS - Use the questions to test your progress. Check your answers on page 94.

1. How does the size of the pupil alter in:
a) Bright light? ....
b) Dim light? ....

2. What is the function of red blood cells?

....

3. Which type of blood cell fights infection and disease?

....

4. What are the two paths that blood takes around the body?

....

5. When is blood oxygenated?

....

6. Where does carbon dioxide get removed from the body?

....

7. Name three products made using fermentation

....

8. What fungus is used for fermentation?

....

9. Water, fibre, vitamins and minerals are all part of a balanced diet; name the other food types.

....

10. What is the difference between arteries and veins?

....

11. Label this diagram of blood:

12. Where does gas exchange take place?

....

13. Where is digested food absorbed into the bloodstream?

....

14. What do we need fibre for?

....

15. What does the central nervous system consist of?

....

16. Without vitamins we can suffer from deficiency diseases. Name a deficiency disease that results from lack of:
a) Vitamin C ....
b) Vitamin D ....

17. How have the alveoli adapted to be an efficient gas exchange surface?

....

....

18. What is the difference between aerobic and anaerobic respiration?

....

19. Name the blood vessels that enter and leave the heart.

20. What is peristalsis?

....

21. Name the three types of enzyme?

....

22. What is the chemical test for protein?

....

23. What type of cells are these and where are they found in the body?

....

24. Describe what happens to the ciliary muscles, suspensory ligaments and lens when you focus on objects close to you.

....

25. Which muscles are involved in breathing?

....

26. Finish the equation for respiration:
Glucose + a) .... ➡ Carbon dioxide + b) ....+ c) ....

27. What role does the liver play in digestion?

....

28. Name the three types of neurone.

....

29. Write out the block diagram of a reflex response.

....

30. What is the equation for anaerobic respiration?

....

How did you do?

1–7	correct	start again
8–15	correct	getting there
16–22	correct	good work
23–30	correct	excellent

CELLS

Cells are the building blocks of life.
All living things are made up of cells.
A living thing is called an organism.
Plants and animals are organisms.

IS IT ALIVE?

To be alive you must have the following characteristics:
Movement (plant leaves move towards the Sun).
Respiration (releasing energy from food).
Sensitivity (responding to changes in the environment).
Growth (to adult size).
Reproduction (producing offspring).
Excretion (getting rid of waste products, such as carbon dioxide).
Nutrition (eating).
• Remember these using <u>MRS</u> <u>GREN</u>; or you could make up your
own way of remembering.

ANIMAL AND PLANT CELLS

They both have:
Nucleus
Cytoplasm
Cell membrane
Mitochondria

Only plant cells have:
Cell wall
Vacuole
Chloroplasts

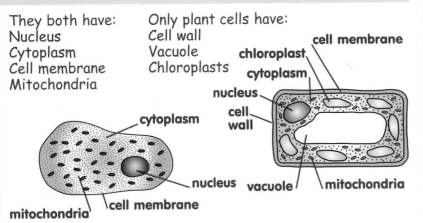

Nucleus – controls all the chemical reactions that take place inside the cell. The nucleus also contains <u>all the information</u> needed to produce a new living organism.
Cytoplasm – where all the <u>chemical reactions</u> take place.
Cell membrane – controls what passes <u>in and out</u> of the cell.
Mitochondria – where <u>respiration</u> takes place. Glucose and oxygen are changed into energy.
Cell wall – made of <u>cellulose</u>, which gives a plant cell <u>strength and support</u>.
Vacuole – contains a weak solution of salt and sugar called <u>cell sap</u>. The vacuole also gives the cell support.
Chloroplasts – contain a green substance called <u>chlorophyll</u>. This absorbs the Sun's energy so that the plant can <u>make its own food</u> during photosynthesis.

CELLS, TISSUES, ORGANS, ORGAN SYSTEMS

· A group of specialised cells working together forms a <u>tissue</u>. A group of muscle cells, for example, is called muscle tissue.
· An <u>organ</u> is made up of tissues working together. The heart, for example, is made of muscle tissue.
· Organs working together make <u>organ systems</u>. The heart and blood vessels work together as part of the circulatory system, to carry food and oxygen around your body.
· Plant cells group together in the same way to form tissues and organs until all the cells together make up an organism.

SPECIAL CELLS

Cells can change their shape in order to carry out a particular job. It's a bit like a factory where each person has their own job. It's more efficient this way. One cell can't do everything.

SPECIALISED ANIMAL CELLS

- *A sperm cell has a <u>tail</u> which enables it to swim towards the egg.*

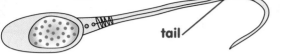

tail

- *Red blood cells have <u>no</u> <u>nucleus</u>, so there is more room for oxygen; they are also <u>biconcave</u> for maximum surface area.*

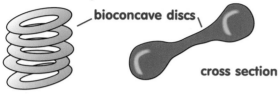

bioconcave discs

cross section

- *Nerve cells are shaped like wires to conduct messages around the body.*

- *Muscle cells can shorten and make a muscle contract and move.*

SPECIALISED PLANT CELLS

- Root hair cells are <u>long</u> and <u>thin</u>, to absorb water and minerals from the soil.

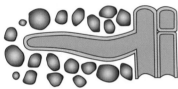

- Xylem are shaped like tubes, to transport <u>water</u> <u>and</u> <u>dissolved</u> <u>minerals</u> up the stem to the leaves.

xylem

- Palisade cells have <u>lots</u> <u>of chloroplasts</u>. They are near the surface of the leaf so they can absorb sunlight for photosynthesis.

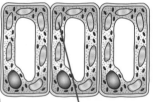

lots of chloroplasts

- Guard cells <u>control the size</u> <u>of the stomata</u> which allow carbon dioxide into the leaf, and oxygen and water out of the leaf. These are on the underside of the leaf.

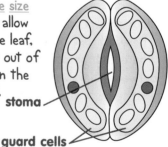

stoma

guard cells

Examiner's Top Tip
Learn at least three examples each of specialised plant and animal cells.

QUICK TEST

1. Name three differences between a plant and an animal cell.
2. Name four similarities between a plant and an animal cell.
3. What does the cell membrane do?
4. What does the cell wall do?
5. What occurs in the mitochondria?
6. What is a specialised cell?
7. A group of specialised cells carrying out the same job is called a _____?
8. How has the sperm cell changed its shape?
9. Which is the only cell not to have a nucleus?
10. Why is it useful that a palisade cell has so many chloroplasts?

10. For photosynthesis; palisade cells are near the surface of the leaf to absorb sunlight
9. Red blood cell
8. Developed a tail
7. Tissue
6. A cell that has changed its shape to do a particular job
5. Respiration
4. Gives a plant cell extra strength and support
3. It controls what passes in and out of the cell
2. Mitochondria, cell membrane, nucleus and cytoplasm
1. Plant cell has chloroplasts, cell wall and a vacuole

DIFFUSION, OSMOSIS AND ACTIVE TRANSPORT

The movement of molecules until they are evenly spread out.

SIMPLE DIFFUSION

The definition of diffusion is:
- The movement of molecules from an area of high concentration to an area of low concentration.
- In other words, it is the natural tendency of molecules to move into all the available space until they are evenly spread out.
- Two rules to remember:

The larger the molecule, the slower the rate of diffusion.

The greater the difference in concentration, the greater the rate of diffusion. The difference is called a concentration gradient.

EXAMPLE OF DIFFUSION IN ANIMAL CELLS
- Your body cells need food and oxygen for respiration.
- These are carried in the blood.
- When the blood reaches the cells, the oxygen and food diffuse into the cells.
- Your cells produce waste and carbon dioxide.
- These diffuse out of the cells into the blood.
- The exchange of carbon dioxide and oxygen between the alveoli in the lungs and the blood is another example of diffusion.

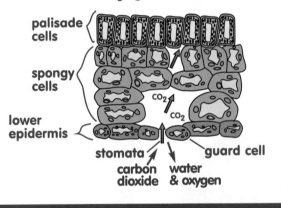

EXAMPLE OF DIFFUSION IN PLANT CELLS
- A plant needs carbon dioxide for photosynthesis.
- Carbon dioxide diffuses into the leaf via the stomata (holes) found on the underside of a leaf.
- A leaf produces oxygen and water vapour.
- Oxygen and water vapour diffuse out of the stomata.
- Diffusion of water vapour occurs much quicker in hot, dry, windy conditions. Think about the best conditions for drying clothes on a line.

ACTIVE TRANSPORT

- Some cells can take up particles against the concentration gradient, i.e. from a low concentration to a high concentration.
- This is called active transport and requires energy from respiration to happen.

AN IMPORTANT OSMOSIS EXPERIMENT

- Visking tubing is a partially permeable membrane and is used to show the effects of osmosis.
- In pure water the visking tubing swells and becomes turgid as water enters by osmosis.
- In strong sugar solution water leaves the visking tubing by osmosis and it becomes flaccid.

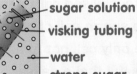

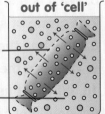

OSMOSIS – A SPECIAL CASE OF DIFFUSION

Each cell is surrounded by a cell membrane, which has tiny holes in it. This membrane is <u>partially</u> <u>permeable</u>. It allows small molecules to pass through, but not larger ones.
The definition of osmosis is:
- The movement of water molecules from an area of high water concentration (weak solution) to an area of low water concentration (strong solution) through a partially permeable membrane.
Water actually moves both ways to try and even up the concentrations. If there is more movement one way, we say there is a <u>net</u> <u>movement</u> of water into the area where there is less water.

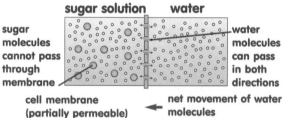

sugar solution   water

sugar molecules cannot pass through membrane

water molecules can pass in both directions

cell membrane (partially permeable)   net movement of water molecules

EXAMPLE OF OSMOSIS IN PLANT CELLS
Root hairs take in water from the soil by <u>osmosis</u>. Water continues to move along the cells of the root and up the xylem to the leaf. All the time water is moving to areas of lower water concentration.

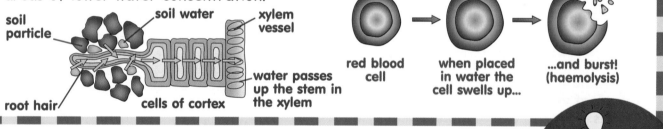

soil particle   soil water   xylem vessel

root hair   cells of cortex   water passes up the stem in the xylem

Osmosis makes plant cells swell up. The water moves into the plant cell vacuole and pushes against the cell wall. The cell wall stops the cell from bursting. We say that the cell is <u>turgid</u>. This is useful as it gives the stem of plants support. If a plant is lacking in water, it wilts and the cells become <u>flaccid</u> as water has moved out of the cell. If a lot of water leaves the cell, the cytoplasm starts to peel away from the cell wall. We say the cell has undergone <u>plasmolysis</u>.

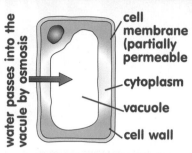

water passes into the vacuole by osmosis

cell membrane (partially permeable)
cytoplasm
vacuole
cell wall

the solution outside the cell is more concentrated than in the vacuole

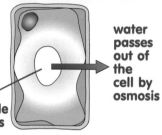

the vacuole shrinks

water passes out of the cell by osmosis

OSMOSIS IN ANIMAL CELLS
Animal cells have no cell wall to stop them swelling. So if they are placed in pure water, they take in water by osmosis until they burst.

red blood cell   when placed in water the cell swells up...   ...and burst! (haemolysis)

QUICK TEST

1. What is diffusion? Can you think of an example?

2. Give an example of diffusion in plant cells.

3. Give an example of where diffusion takes place in an animal.

4. The greater the concentration difference the _____ the diffusion rate?

5. What is a partially permeable membrane?

6. What substance moves by osmosis?

7. What happens to plant cells that take up water by osmosis?

8. What happens to plant cells that lose water by osmosis?

9. Where does the energy come from for active transport?

10. Define active transport.

1. The movement of molecules from an area of high concentration to an area of low concentration, e.g. smell of a match, cooking, etc.
2. Diffusion of carbon dioxide into a leaf
3. Alveoli in the lungs
4. Faster
5. A membrane that only allows small molecules through
6. Water
7. They become turgid
8. They become flaccid
9. Respiration
10. Taking up particles against the concentration gradient

PLANT AND LEAF – STRUCTURE AND FUNCTION

THE LEAF

waxy cuticle

epidermis

palisade cell

spongy layer

leaf vein

guard cell

stoma

The leaf is the **organ of photosynthesis**. It makes all the food for the plant.

- The upper surface of the leaf is called the **waxy cuticle**. It is a waterproof layer that cuts down the loss of water by evaporation.
- The upper cells of the leaf make up the **epidermis**. Light passes straight through these.
- The next layer of cells contains the **palisade cells**. This is where most photosynthesis takes place.
- These contain lots of **chloroplasts**. The chloroplasts contain a pigment called chlorophyll. **Chlorophyll** absorbs sunlight for photosynthesis.
- The **spongy layer** contains rounded cells with lots of air spaces. This allows carbon dioxide to circulate and reach the palisade cells.
- The **leaf vein** contains the **xylem and phloem** tubes. They run over the plant, supplying it with water and taking away the glucose that is produced.
- At the lower surface of the leaf are tiny pores called **stomata** (one pore = one stoma). The stomata open and close to let carbon dioxide in and water vapour and oxygen out.

- **Guard cells** surround the stomata and control their opening and closing.
- When water is in short supply, the guard cells become **flaccid** and less curved; this closes the stomata and prevents water being lost from the leaf.
- When a plant has plenty of water the guard cells become **turgid** and curved; the stomata are open and water can escape from the leaf.
- The leaf has many features that enable it to carry out photosynthesis in the best possible way. These include:
- Leaves are **flat with a large surface area** to absorb as much sunlight as possible.
- They are **thin**, so carbon dioxide can reach the inner cells easily.
- They have **plenty of stomata** in the lower skin.
- They have **plenty of veins** to support the leaf and carry substances to and from all the cells in the leaf and plant.

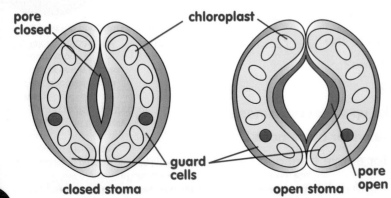

pore closed

chloroplast

guard cells

closed stoma

open stoma

pore open

The plant's basic structure is divided up into <u>five parts</u>.
The parts of a plant have adapted to do a particular job or function.

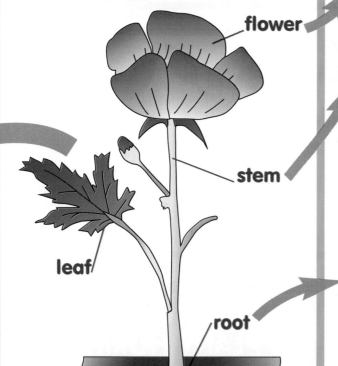

flower

stem

leaf

root

root hair

THE FLOWER

This contains the male and female sex organs. These make seeds.
The flower is usually brightly coloured to attract insects for pollination.

THE STEM

This holds the plant <u>upright</u>.
It contains hollow tubes called <u>xylem</u> and <u>phloem</u>.
Xylem tubes <u>carry</u> <u>water</u> and <u>dissolved</u> <u>minerals</u> from the roots to the leaves.
Phloem <u>carries</u> <u>glucose</u> made by the leaf in photosynthesis up and down the plant.

THE ROOT

The root's main job is anchoring the plant in the soil.
They also take up <u>water</u> <u>and</u> <u>minerals</u> from the soil.
Remember: Minerals are not food.

THE ROOT HAIRS

The actual place where water and minerals are absorbed from the soil.
Root hairs <u>increase</u> <u>the</u> <u>surface</u> <u>area</u> of the roots for more efficient absorption.

Examiner's Top Tip
Learn the five parts of a flowering plant and what they are for.

Examiner's Top Tip
The adaptations of the leaf for photosynthesis are important. Learn what makes the leaf so good at its job.

QUICK TEST

1. What is the job of the leaf?
2. Name the part of the plant that keeps it upright.
3. What do the guard cells do?
4. In what situation do the guard cells close the stomata?
5. When are the guard cells turgid and the stomata open?
6. In what structure do water and minerals travel up the stem?
7. The _____ carries glucose up and down the plant.
8. What is the root's main job?
9. What do the palisade cells contain a lot of?
10. What are the air spaces for in the spongy layer?

10. Allow circulation of carbon dioxide
9. Chloroplasts/chlorophyll
8. Anchor the plant
7. Phloem
6. Xylem cells
5. When water is plentiful
4. Shortage of water
3. Control the opening and closing of the stomata
2. Stem
1. Photosynthesis

CONTROL OF GROWTH

- *Plant growth takes place mainly in the <u>root tip</u> and <u>shoot tip</u>.*
- *The root tip and shoot tip contain a hormone called <u>auxin</u>.*
- *Auxin <u>speeds up growth in stems</u> and <u>slows down growth in roots</u>.*

Examiner's Top Tip
You may be asked to describe ways that plant hormones are used for commercial purposes. Learn the four examples.

RESPONSE TO WATER

- A plant's response to water is called <u>hydrotropism</u>.
- Roots always grow to a certain extent towards water, even if it means ignoring the pull of gravity and growing sideways.
- An uneven amount of moisture will cause more auxin to appear on the side with more water.
- This inhibits the growth of cells on this side.
- The root cells on the outside will grow quicker and will bend towards the moisture.

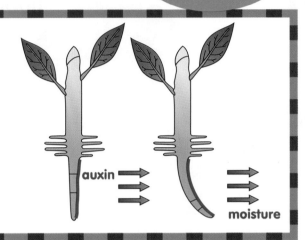

RESPONSE TO GRAVITY

- A plant's response to gravity is called **geotropism**.
- Even if you plant a seed the wrong way up, the shoot always grows up, away from gravity and the root grows down towards gravity.
- If a plant is put on its side, auxin gathers on the lower half of the shoot and root.

auxin <u>slows</u> down the growth of root cells, so the root curves downwards.

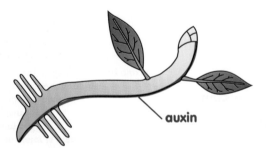

auxin <u>speeds</u> up the growth of shoot cells so the shoot curves upwards.

COMMERCIAL USES OF PLANT HORMONES

1. GROWING CUTTINGS
- Rooting powder contains synthetic auxins.
- A cutting is taken from a plant and dipped in this powder.
- This stimulates the roots to grow quickly and enables gardeners to grow lots of exact copies of a particular plant.

2. KILLING WEEDS
- <u>Synthetic auxins</u> are used as selective weedkillers.
- They only affect the broad-leaved weeds; narrow-leaved grasses and cereals are not affected.
- They kill the weed by making the weed grow too fast.

3. SEEDLESS FRUITS
- <u>Synthetic auxins</u> are sprayed on unpollinated flowers.
- Fruits form without fertilisation.
- These fruits form without pips, e.g. seedless grapes.

4. EARLY RIPENING
- Plant hormones can also be used to ripen fruit in transport.
- Bananas are picked when they are unripe and less easily damaged.
- By the time they arrive for sale they are yellow and not green.

PLANT SENSES

- Plants respond to their surroundings to give them a better chance of survival.
- Plant responses are called <u>tropisms</u> and are controlled by a <u>hormone</u>.
- Plants respond to <u>light</u>, <u>gravity</u> <u>and</u> <u>water</u>.

Examiner's Top Tip
Remember: unequal distribution of auxin speeds up growth in shoots and slows down growth in roots.

shoot — hormone gathers on lower side and increases growth in the shoot – it curves upwards

seed

root

hormone collects on lower side and decreases growth in the root – it curves downwards

RESPONSE TO LIGHT

- A plant's response to light is called <u>phototropism</u>.
- Plants need light to make food during photosynthesis.
- Plants will grow towards the light.

Normally light shines from above. Auxin is spread evenly and the shoot grows upwards.

auxin is made here
light
auxin moves down the stem

If light comes from one side, auxin accumulates down the shaded side. Auxin makes these cells grow faster.

light
auxin

The result is that the shoot <u>bends</u> <u>towards</u> <u>the</u> <u>light</u>.

auxin makes cells grow faster here
light

QUICK TEST

1. What is the name of the hormone that controls plants responses?

2. A plant's response to gravity is called _____?

3. A plant's response to light is called _____?

4. In which parts of a plant is auxin made?

5. If a light shines from the left.
   onto a plant, on which side does auxin gather?

6. Auxin speeds up/slows down growth in roots?

7. Auxin speeds up/slows down growth in shoots?

8. Name four ways in which plant hormones can be used to benefit gardeners.

9. How does a shoot know which way to grow if it is laid on its side?

10. How does a root know which way to grow if the seed is planted upside down?

10. Auxin gathers on the lower half of the root and slows down growth on this shoot
9. Auxin gathers on the lower half of the shoot and speeds up the growth on this side
8. Growing cuttings; killing weeds; seedless fruits; early ripening
7. Speeds up
6. Slows down
5. Opposite side (right)
4. Root tips and shoot tips
3. Phototropism
2. Geotropism
1. Auxin

PLANT TRANSPORT AND GOOD HEALTH

Plants have separate transport systems:
- <u>Xylem</u> <u>tissue</u> transports water and minerals from the roots to the stem and leaves.
- <u>Phloem</u> <u>tissue</u> carries nutrients from the leaves to the rest of the plant.
- <u>Transpiration</u> is the movement of water through the plant. It begins in the roots and ends in the leaves, where it is lost by evaporation through the stomata on the underside of leaf.

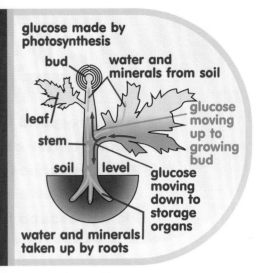

glucose made by photosynthesis

bud

water and minerals from soil

leaf

stem

soil level

glucose moving up to growing bud

glucose moving down to storage organs

water and minerals taken up by roots

TRANSPORT OF FOOD

Sugars are made in the leaves by photosynthesis.
The sugars are converted to <u>starch</u> and are then transported by the phloem.
The <u>phloem</u> <u>vessels</u> run up and down the plant from the <u>leaves</u> to the <u>storage</u> <u>organs</u> and <u>growing</u> <u>parts</u> of the plant.
They transport sugars and other food substances made by the plant cells, such as amino acids and fatty acids.
Phloem and xylem vessels often run side by side and form a <u>vascular</u> <u>bundle</u>.
Vascular bundles are found in the roots, stem and veins of a leaf.

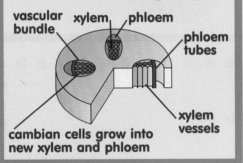

vascular bundle   xylem   phloem

phloem tubes

xylem vessels

cambian cells grow into new xylem and phloem

HEALTHY GROWTH

There are three essential minerals needed for growth (nitrates, phosphates and potassium). Minerals are obtained from the soil, (dissolved in water), by active transport.
- <u>Nitrates</u> are needed for making amino acids, proteins and DNA.
- <u>Phosphates</u> play an important role in photosynthesis and respiration. Phosphorus is also used for making DNA and cell membranes.
- <u>Potassium</u> is involved in making the enzymes used in respiration and photosynthesis work.
- <u>Magnesium</u> and <u>iron</u> are also needed in small amounts to make chlorophyll.

Lack of nutrients causes the following mineral deficiency symptoms:

lack of nitrates causes stunted growth and yellow, older leaves

lack of phosphates causes poor root growth and purple, young leaves

lack of potassium causes yellow leaves with dead spots

WATER

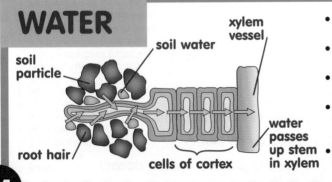

soil water

xylem vessel

soil particle

root hair

cells of cortex

water passes up stem in xylem

- <u>Water</u> <u>and</u> <u>dissolved</u> <u>minerals</u> enter the root cells by <u>osmosis</u>.
- The <u>root</u> <u>hairs</u> on the surface of the root provide a <u>large</u> <u>surface</u> <u>area</u> for absorption.
- Osmosis causes the water to move from cell to cell until it reaches the xylem vessels.
- The xylem are dead cells joined together to make up tubes. They have thick strong walls made of <u>lignin</u>, which gives the plant support.
- The water moves up the stem to the leaves in the <u>transpiration</u> <u>stream</u>.

FACTORS AFFECTING TRANSPIRATION

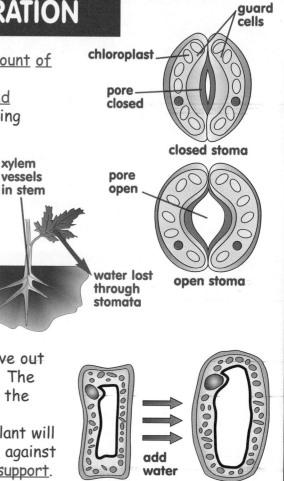

chloroplast

guard cells

pore closed

closed stoma

pore open

open stoma

xylem vessels in stem

water lost through stomata

1. The rate of transpiration is affected by the <u>amount of light</u>, <u>temperature</u>, <u>air movement</u> and <u>humidity</u>.
2. Transpiration is fastest when it is <u>hot</u>, <u>dry and windy</u>; think about the best conditions for drying clothes on a line.
3. Most transpiration takes place during the day and not at night; this is because the stomata are closed at night and open in the day.
4. <u>Guard cells</u> control the <u>opening and closing of the stomata</u>.
5. The stomata may also be closed in very dry conditions to reduce transpiration.
6. Plants that live in hot, dry climates have <u>less stomata</u> and also have a <u>thick</u>, <u>waxy cuticle</u> to reduce evaporation from the leaves' surface.
7. If there is a lack of water, then water will move out of the plant cells and they will become <u>flaccid</u>. The stem can no longer keep the plant upright, and the plant will <u>wilt</u>.
8. If there is enough water then the cells of the plant will draw in water. The contents of the cell will push against the cell wall, making it <u>turgid</u>. This gives the plant <u>support</u>.

add water

flaccid cell   **turgid cell**

THE TRANSPIRATION STREAM

- The <u>loss of water</u> from the plant's leaves by evaporation is out of the stomata and is called <u>transpiration</u>.
- The flow of water up the xylem to the leaves is called the <u>transpiration stream</u>.
- As water is used up during photosynthesis or lost from the leaves by evaporation, more is drawn up through the roots.
- The transpiration stream also draws minerals into the plant.

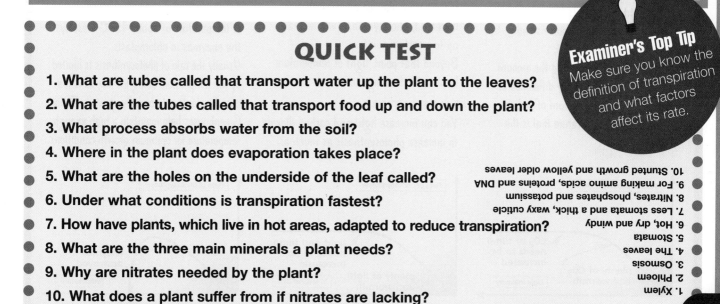

QUICK TEST

1. What are tubes called that transport water up the plant to the leaves?
2. What are the tubes called that transport food up and down the plant?
3. What process absorbs water from the soil?
4. Where in the plant does evaporation takes place?
5. What are the holes on the underside of the leaf called?
6. Under what conditions is transpiration fastest?
7. How have plants, which live in hot areas, adapted to reduce transpiration?
8. What are the three main minerals a plant needs?
9. Why are nitrates needed by the plant?
10. What does a plant suffer from if nitrates are lacking?

10. Stunted growth and yellow older leaves
9. For making amino acids, proteins and DNA
8. Nitrates, phosphates and potassium
7. Less stomata and a thick, waxy cuticle
6. Hot, dry and windy
5. Stomata
4. The leaves
3. Osmosis
2. Phloem
1. Xylem

PHOTOSYNTHESIS

- Photosynthesis is a chemical process that plants use to make their food (glucose) using energy from the Sun. It occurs in the leaves.
- Photosynthesis occurs in the light; respiration occurs all of the time.

The word equation of photosynthesis:

$$\text{carbon dioxide + water} \xrightarrow[\text{chlorophyll}]{\text{light}} \text{glucose + oxygen}$$

The balanced symbol equation is:

$$6CO_2 + 6H_2O \Rightarrow C_6H_{12}O_6 + 6O_2$$

THE LEAF – THE ORGAN OF PHOTOSYNTHESIS

1. <u>Carbon dioxide</u> enters the leaf through tiny holes on the underside of the leaf by diffusion. The holes are called <u>stomata</u>.
2. The <u>spongy layer</u> of cells has air spaces for the carbon dioxide to circulate and for the exchange of gases.
- <u>Oxygen</u> produced by photosynthesis diffuses out through the stomata (some is retained for respiration).
3. <u>Chloroplasts</u> are most abundant on the upper surface of the leaf in <u>palisade cells</u>. Chloroplasts contain <u>chlorophyll</u>.
- <u>Chlorophyll</u> is a green pigment that <u>absorbs sunlight</u> energy.

4. Inside the leaf are <u>veins</u>; these are continuous with the stem and roots of the plant.
- The veins contain <u>xylem</u> and <u>phloem</u>.
- <u>Xylem</u> transports <u>water</u> from the roots to the leaves.
- <u>Phloem</u> transports the <u>glucose</u> up and down the plant from the leaves, to where it is needed, particularly the growing regions (the bud) and the storage areas (the roots).

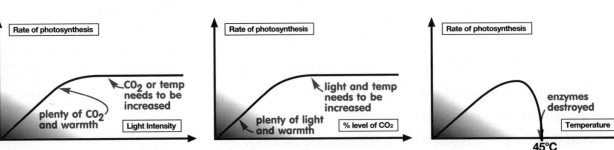

FACTORS AFFECTING THE RATE OF PHOTOSYNTHESIS

We can measure the rate of photosynthesis by how much oxygen is produced in a given time. There are three things that affect the rate of photosynthesis. We call them <u>limiting factors.</u> They are:

LIGHT

- If the light intensity is increased, photosynthesis will increase steadily, but only up to a certain point.
- After this point, increasing the amount of light will not make any difference as it will be either the amount of carbon dioxide or the temperature that is the limiting factor.

CARBON DIOXIDE

- If the carbon dioxide concentration is increased, photosynthesis will increase up to a certain point.
- Beyond that point, light or temperature become the limiting factor.

TEMPERATURE

- You can increase light and carbon dioxide to increase photosynthesis as much as

possible, but the temperature must not get too hot or too cold.
- A temperature of about 45^0C destroys the enzymes in chloroplasts.
- Usually the rate of photosynthesis is limited by the temperature being too low, as is the case for plants not normally grown in Britain.
- Greenhouses help maintain a high enough temperature for optimum growth conditions.

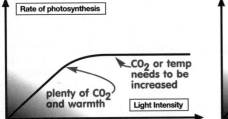

PHOTOSYNTHESIS EXPERIMENTS

- We can prove that a plant needs chlorophyll, light and carbon dioxide for photosynthesis.
- We can also prove that a plant produces oxygen and glucose, which it changes into starch.

- Test the leaf for starch; only the uncovered part of the leaf will contain starch.

aluminium foil

starch present

no starch

IS STARCH PRODUCED IN PHOTOSYNTHESIS?

1. Dip a leaf in boiling water for about a minute to soften it.
2. Put the leaf in a test-tube of ethanol and stand in hot water for 10 minutes (this removes the colour).
3. Remove and wash the leaf.
4. Lay the leaf flat in a petri dish and add iodine.
5. If starch is present the leaf should go blue/black.
- This experiment can also prove that chlorophyll is needed if you use a variegated leaf. A variagated leaf has some white parts where there is no chlorophyll.
- The plant should only go black where the leaf was green.

iodine

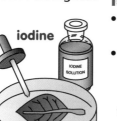

IODINE SOLUTION

DOES A PLANT NEED LIGHT?

- De-starch a plant by leaving it in the dark for 24 hours.
- Cover part of leaf with some foil and leave in the sun for a few hours.

DOES A PLANT NEED CARBON DIOXIDE?

- De-starch a plant; enclose it in a clear bag containing sodium hydroxide (this absorbs carbon dioxide).
- After a few hours, test a leaf for starch; none should be present.

plastic bag

elastic band

soda lime

IS OXYGEN PRODUCED IN PHOTOSYNTHESIS?

- Set up the apparatus as shown in the diagram and collect the bubbles of gas.
- Test the gas for oxygen using a glowing splint. It should relight.

inverted funnel

light

weak hydrogen carbonate solution (gives CO_2 to plant)

pond weed

plasticine supports

USES OF GLUCOSE

- Some glucose is used in respiration to obtain energy. Other uses include converting it to:
- Insoluble starch stored in the roots, particularly in the winter. In this form it does not cause too much water to move into the cells by osmosis, as it doesn't contribute to the concentration inside the cells.

- Cellulose, needed for cell walls.
- Lipids and oils are formed from glucose and stored in seeds.
- Glucose can also be combined with other substances, such as nitrates obtained from the soil and turned into proteins.

Examiner's Top Tip
The three graphs of limiting factors are important to learn; copy them down and in each case state the limiting factors.

QUICK TEST

1. What does a plant need for photosynthesis?
2. What does a plant produce in photosynthesis?
3. List three uses of the glucose produced.
4. What three factors limit the rate of photosynthesis?
5. By what process does the plant generate energy?
6. What is the advantage of storing glucose as starch?
7. How does the plant obtain water?
8. How does the plant obtain carbon dioxide?
9. What is chlorophyll?
10. Why do you think that the palisade cells are near the surface of the leaf?

10. So they can absorb as much sunlight as possible through their chloroplasts
9. A green pigment that absorbs sunlight
8. From the air, through the stomata
7. Through the roots and xylem tubes up to the leaf
6. Prevents the cells taking in too much water by osmosis
5. Respiration
4. Amount of light, carbon dioxide and temperature
3. Cellulose, starch, protein, respiration, etc
2. Oxygen and glucose
1. Carbon dioxide, water, chlorophyll and light

THE SKIN

- Your skin has many important functions.
- It is the first line of defence against germs.
- It contains <u>receptors</u> that are sensitive to touch, pain, temperature and pressure that relay these messages to the brain.
- It helps to <u>keep your body temperature constant</u>.

outer layer
made of dead flat cells that flake off; they are replaced by cells from below and provide a barrier to the outside environment

hair

sweat pore

blood capillary

sweat glands produce liquid that is secreted at the surface to cool the body down

hair root the living part of the hair contained in the hair follicle

fat layer for insulation

blood vessel

sebaceous glands produce sebum, an oily substance that makes the skin waterproof and supple

HYPOTHERMIA

- *Despite our body's attempts at maintaining a constant internal temperature, things can go wrong.*
- *If the outside temperature is extremely cold and the body temperature falls dramatically, the control centre in the brain stops working.*
- *The body temperature gets lower and lower and sufferers may slip into a coma and die if no action is taken.*
- *Babies and old people are particularly susceptible to hypothermia.*

WHEN IT IS HOT

<u>Blood</u> <u>vessels</u> at the surface of the skin <u>widen</u>. This allows more blood to flow to the surface. This is called <u>vasodilation</u>.
Heat is lost as it <u>radiates</u> from the skin. You look flushed.
Sweat glands begin to secrete sweat. The sweat <u>evaporates</u> from the skin and takes away <u>heat</u> <u>energy</u>.
In animals that do not sweat <u>panting</u> helps them to keep cool.

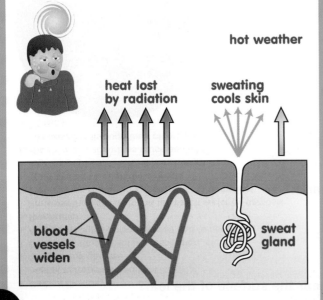

hot weather

heat lost by radiation

sweating cools skin

blood vessels widen

sweat gland

WHEN IT IS COLD

Many warm-blooded animals have a thick layer of fat beneath their skin. This helps <u>insulate</u> our bodies in cold weather. Our skin has other mechanisms to deal with the cold:

- <u>**Blood vessels**</u> – at the surface of the skin <u>**contract**</u>, so that very little blood gets to the surface. This is called <u>**vaso-constriction**</u>.
- Very little heat is lost by <u>**radiation**</u>.
- You look pale.
- <u>**Shivering**</u> – your muscles contract quickly, which produces extra heat to warm your body.
- <u>**Sweat glands**</u> – stop producing sweat.

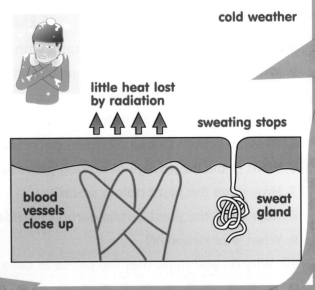

cold weather

little heat lost by radiation

sweating stops

blood vessels close up

sweat gland

SKIN AND BODY TEMPERATURE

· We are warm-blooded and have mechanisms that can keep our body temperature constant.
· Our body's reaction to changes in temperature is controlled by the hypothalamus in the brain.
· Inside your body the temperature always stays around the same temperature, 37°C.
· This is because the enzymes involved in the body's chemical reactions work best at 37°C.
· Your skin helps to maintain this temperature (homeostasis).

BEHAVIOUR

• When it is hot, the receptors in our skin send messages to the control centre in the brain so that we can feel the changes in temperature.
• This enables humans to change their behaviour to keep warm. We put on extra clothes and move around more.
• When it is hot we can take off unnecessary clothes and stay in the shade.
• Cold-blooded animals do not have the same level of control as warm-blooded animals. So they have to change their behaviour in different ways to survive.
• For example, lizards must stay in the sun to keep warm or go in the shade to cool down.

BODY HAIR

• Many animals have a thick layer of hair or fur.
• In cold weather the hair stands up.
• This traps a layer of air close to the skin.
• Air insulates the body and prevents heat loss.
• As humans, who have much less hair, this is not much use. But it still happens and we get goose bumps.

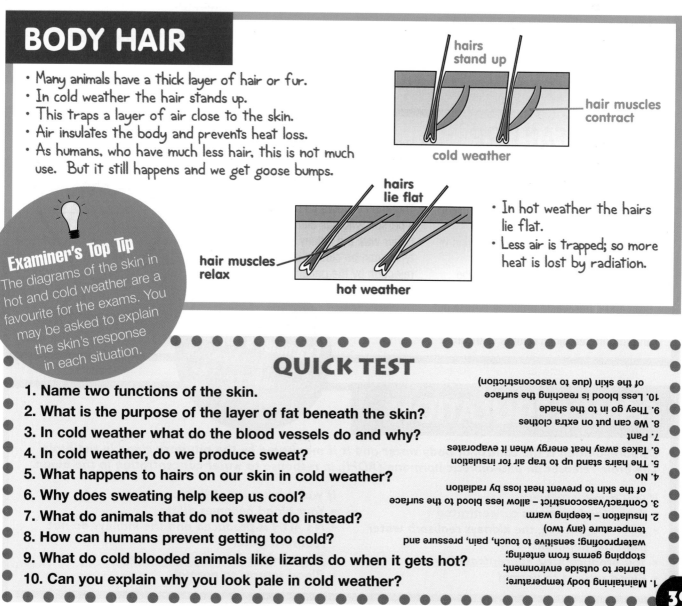

hairs stand up
hair muscles contract
cold weather

hairs lie flat
hair muscles relax
hot weather

• In hot weather the hairs lie flat.
• Less air is trapped; so more heat is lost by radiation.

Examiner's Top Tip
The diagrams of the skin in hot and cold weather are a favourite for the exams. You may be asked to explain the skin's response in each situation.

QUICK TEST

1. Name two functions of the skin.
2. What is the purpose of the layer of fat beneath the skin?
3. In cold weather what do the blood vessels do and why?
4. In cold weather, do we produce sweat?
5. What happens to hairs on our skin in cold weather?
6. Why does sweating help keep us cool?
7. What do animals that do not sweat do instead?
8. How can humans prevent getting too cold?
9. What do cold blooded animals like lizards do when it gets hot?
10. Can you explain why you look pale in cold weather?

1. Maintaining body temperature; barrier to outside environment; stopping germs from entering; waterproofing; sensitive to touch, pain, pressure and temperature (any two)
2. Insulation – keeping warm
3. Contract/vasoconstrict – allow less blood to the surface of the skin to prevent heat loss by radiation
4. No
5. The hairs stand up to trap air for insulation
6. Takes away heat energy when it evaporates
7. Pant
8. We can put on extra clothes
9. They go in to the shade
10. Less blood is reaching the surface of the skin (due to vasoconstriction)

THE KIDNEYS

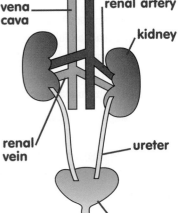

- The kidneys are organs of <u>excretion</u>; they remove the waste products urea, excess water and ions, such as salt.
- They first <u>filter</u> the blood, then <u>reabsorb</u> what the body needs.
- There are two kidneys situated towards the back of the body, just above the waist.

HOMEOSTASIS

- Homeostasis is the mechanism by which the body maintains a <u>constant</u> <u>internal</u> <u>environment</u>.
- Examples of homeostasis include keeping the body temperature constant and controlling blood sugar levels.
- Information is passed to the brain, and the brain sends messages back to adjust the levels back to normal.
- The kidney has a major role in homeostasis. It controls the amount of water in our body as well as the removal of excess substances and the poisonous substance, urea.

KIDNEY FAILURE

- You can survive with one kidney, but if both fail, <u>dialysis</u> is needed.
- Dialysis involves filtering the blood using a machine.
- The machine is connected up to a vein in the patient's arm.
- The blood then passes through a tube in the machine.
- Urea and other waste products diffuse out of the blood through a dialysis membrane and the filtered blood returns to the arm.

- Sugar and ions that are needed do not diffuse out of the blood into the dialysis fluid.
- A person with failed kidneys may have a <u>kidney</u> <u>transplant</u>.
- The kidney would be removed from either a person who has died or a living donor. Either way, the kidney's tissue and blood type must be closely matched to prevent rejection by the patient.

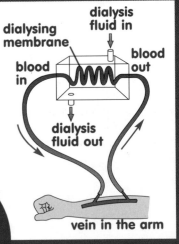

Examiner's Top Tip
Learn the diagram of the kidney and its associated organs.

OSMOREGULATION

Osmoregulation is the control of body water and it is monitored by the pituitary gland in the brain. The brain releases an anti-diuretic hormone (ADH), in response to water concentration in the blood.

If you drink <u>less</u> <u>fluids</u>:
- *Your blood becomes <u>concentrated</u>*
- *<u>ADH</u> <u>is</u> <u>produced</u>; the kidneys reabsorb water from the nephrons*
- *You produce little <u>concentrated</u> <u>urine</u>*
- *Blood returns to normal*

If you drink a <u>lot</u> <u>of</u> <u>fluids</u>:
- *Your blood becomes <u>dilute</u>*
- *<u>No</u> <u>ADH</u> is produced and the kidneys do not reabsorb water*
- *You produce lots of <u>dilute</u> <u>urine</u>*
- *Blood returns to normal*

ULTRAFILTRATION

- The blood arrives at the kidney in the <u>renal artery</u> <u>at</u> <u>high</u> <u>pressure</u> and enters the group of capillaries called the <u>glomerulus</u>.
- This high pressure squeezes water, urea, ions and glucose out of the blood into the <u>bowman's capsule</u>. Large molecules stay in the blood.
- Liquid in the <u>nephron</u> contains useful substances like glucose and some ions.

FUNCTION OF THE KIDNEY

- <u>**The kidneys' role is to filter the blood.**</u>
- **They remove urea and adjust ion and water content.**

REMOVE UREA
Urea is a poisonous substance produced in the liver from excess amino acids.

ADJUST ION CONTENT
Ions are taken into the body in food and are absorbed into the blood, any excess ions are removed by the kidneys and in sweat.

ADJUST WATER CONTENT
Water is taken in through food and drink, and removed in faeces and breath. Excess water is removed in the kidneys and makes up urine. Excess water is also removed in sweat.

PARTS OF THE KIDNEY

REABSORPTION AND RELEASE OF WASTE SUBSTANCES

- *As the blood flows along the nephron, useful substances are <u>reabsorbed</u> <u>by</u> <u>active</u> <u>transport</u> <u>using</u> <u>energy</u>.*
- *The cleaned blood enters the renal vein and leaves the kidney.*
- *How much water is reabsorbed depends on a hormone called <u>ADH</u> (antidiuretic hormone).*
- *All urea and excess ions and water are not reabsorbed and pass into the collecting duct.*
- *This fluid continues out of the nephron into the ureter and down to the bladder as <u>urine</u>.*

QUICK TEST

1. Where are your kidneys situated?
2. Define homeostasis.
3. What is the kidneys' main role?
4. What are the two stages of cleaning the blood?
5. What is osmoregulation?
6. What hormone controls the amount of water in the blood?
7. What happens if you do not drink a lot of fluids?
8. Which substances are reabsorbed?
9. What is urea?
10. How does the cleaned blood leave the kidneys?

Examiner's Top Tip
Follow the diagram of the kidney nephron as you read the description of what happens at each stage.

1. At the back, above the waist
2. Maintaining a constant internal environment
3. Cleaning the blood, excretion
4. Ultrafiltration and reabsorption
5. Regulation of blood water concentration
6. ADH
7. ADH is released, kidney reabsorbs more water
8. Glucose, some ions and some water
9. A poisonous substance formed from excess amino acids
10. Via the renal vein

CAUSES OF DISEASE

- **Microbes are bacteria, fungi and viruses.**
- **Not all microbes cause disease; some are useful.**
- **Microbes that get inside you and make you feel ill are called pathogens or germs.**
- **Pathogens rapidly reproduce in warm conditions when there is plenty of food.**

HOW ARE DISEASES SPREAD?

- **Contact** with infected people, animals or objects used by infected people; e.g. athlete's foot, chicken pox and measles.
- Through the **air**, e.g. flu, colds and pneumonia.
- Through infected **food and drink**, e.g. cholera from infected drinking water and salmonella food poisoning.

air

food

touch

water

Examiner's Top Tip
Remember that not all microbes are harmful and cause disease.

FUNGI

- *Fungi cause diseases such as athlete's foot and ringworm.*
- *Fungi reproduce by making spores that can be carried from person to person.*
- *Most fungi are useful as decomposers. Yeast is a fungus that is used when making bread, beer and wine.*

BACTERIA

Bacteria are living organisms that feed, move and carry out respiration.
- There are three main shapes of bacteria:

rods (bacilli)   spheres (cocci)   spirals (spirilla)

some bacteria have circular DNA called plasmids; these are useful in genetic engineering. You will read about it later on in the book

cell wall

cell membrane

cytoplasm

bacterial cells have no nucleus but do have genes in the cytoplasm

Bacteria reproduce rapidly by mitosis and produce exact copies of themselves.
Bacteria are good at surviving in unfavourable conditions such as extreme temperatures. They form a protective coat around themselves and are then called spores.
When conditions return to normal, the bacterial cell comes out of its spore and continues to reproduce.

HOW THEY CAUSE DISEASE

- Bacteria destroy living tissue. For example, tuberculosis destroys lung tissue.
- Bacteria can produce poisons, called toxins. For example food poisoning is caused by bacteria releasing toxins.

VIRUSES

Viruses consist of a <u>protein</u> <u>coat</u> surrounding a few <u>genes</u>.

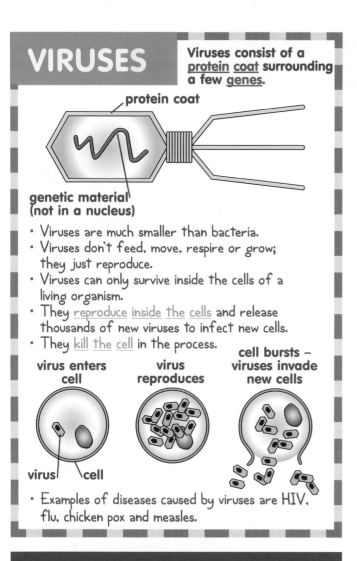

protein coat

genetic material (not in a nucleus)

- Viruses are much smaller than bacteria.
- Viruses don't feed, move, respire or grow; they just reproduce.
- Viruses can only survive inside the cells of a living organism.
- They <u>reproduce</u> <u>inside</u> <u>the</u> <u>cells</u> and release thousands of new viruses to infect new cells.
- They <u>kill</u> <u>the</u> <u>cell</u> in the process.

virus enters cell

virus reproduces

cell bursts – viruses invade new cells

virus   cell

- Examples of diseases caused by viruses are HIV, flu, chicken pox and measles.

HOW DO PATHOGENS GET IN?

- Pathogens have to enter our body before they can do any harm.

respiratory systems – droplets of moisture containing viruses are breathed in

digestive system – pathogens get in via food and drink

skin – if the skin is <u>damaged</u>, pathogens can get in

reproductive system – diseases can be passed on through sexual intercourse

SYMPTOMS OF INFECTION

- Symptoms are the effects diseases have on the body; they are usually caused by the toxins released by the pathogens.
- Symptoms include a high temperature, headache, loss of appetite and sickness.

VECTORS

Some pathogens rely on <u>vectors</u> to transfer them from one organism to another. A vector is an organism that transports a pathogen. An example would be a <u>mosquito</u>.

A mosquito carrying the parasite that causes malaria may infect another person by injecting the parasite into the person's bloodstream when it bites them.

Examiner's Top Tip
Learn the structure of a bacterium and a virus; notice the similarities and differences between these and general animal and plant cells.

QUICK TEST

1. Name the three types of microbe.

2. What do we call microbes that cause disease?

3. How do bacteria cause disease?

4. How do viruses cause disease?

5. Give two examples of diseases caused by fungi.

6. How are infections spread?

7. What is a vector? Give an example.

8. Name three ways in which pathogens can enter the body.

9. Name two diseases caused by viruses.

10. Name two diseases caused by bacteria.

10. Food poisoning, tuberculosis, cholera
9. HIV, colds, flu, chicken pox and measles
8. Via skin, digestive, reproductive and respiratory systems, and vectors
7. An organism that transports a diseases from person to person, e.g. mosquito
6. By contact, air and food and drink
5. Athlete's foot and ringworm
4. They reproduce inside living cells and kill them
3. Destroy living tissue and produce toxins
2. Pathogens or germs
1. Virus, bacteria and fungi

PREVENTION IS BETTER THAN CURE

The human body has several ways of preventing disease-causing microbes from entering; these are called natural defences.

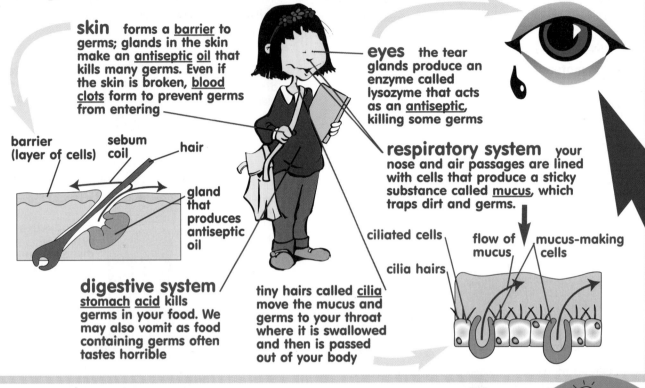

skin forms a <u>barrier</u> to germs; glands in the skin make an <u>antiseptic</u> <u>oil</u> that kills many germs. Even if the skin is broken, <u>blood clots</u> form to prevent germs from entering

barrier (layer of cells)   sebum coil   hair

gland that produces antiseptic oil

eyes the tear glands produce an enzyme called lysozyme that acts as an <u>antiseptic</u>, killing some germs

respiratory system your nose and air passages are lined with cells that produce a sticky substance called <u>mucus</u>, which traps dirt and germs.

ciliated cells   flow of mucus   mucus-making cells

cilia hairs

digestive system <u>stomach</u> <u>acid</u> kills germs in your food. We may also vomit as food containing germs often tastes horrible

tiny hairs called <u>cilia</u> move the mucus and germs to your throat where it is swallowed and then is passed out of your body

ANTIBIOTICS

- Sometimes bacteria get through the body's defences and reproduce successfully.
- In this case outside help in the form of <u>antibiotics</u> is needed to kill the germs.
- Antibiotics kill the germs without harming the body cells.
- <u>Penicillin</u> was the first form of antibiotic. It is made from a mould called Penicillium notatum.
- <u>Antibiotics</u> <u>cannot</u> <u>treat</u> <u>infections</u> <u>caused</u> <u>by</u> <u>viruses</u>. The body has to fight them on its own.
- Antibiotics can kill most bacteria, but as we continue to use them, bacteria are becoming <u>resistant</u> to them.
- New antibiotics are constantly needed to fight the battle against bacteria.
- There are no drugs to kill viruses. You just have to wait for your body to deal with them.

THE IMMUNE SYSTEM RESPONSE

If pathogens get into your body, <u>white</u> <u>blood</u> <u>cell</u>s travelling around in your blood spring into action.
- White blood cells can make chemicals called <u>antitoxins</u> that destroy the toxins produced by bacteria.
- White blood cells called <u>phagocytes</u> try to engulf bacteria or viruses before they have a chance to do any harm.
- However, if the pathogens are in large numbers, the other type of white blood cell, called <u>lymphocytes</u>, are involved.

All germs have chemicals on their surface called antigens.

Lymphocytes recognise these antigens as foreign.

<u>Lymphocytes</u> <u>produce</u> <u>chemicals</u> <u>called</u> <u>antibodies</u> that attach to these antigens and clump them together. <u>Phagocytes</u> can then engulf and destroy the bacteria and viruses.

PREVENTING THE SPREAD OF GERMS

- We can <u>sterilise</u> equipment used in food preparation or in operating theatres by heating it to a temperature of 120°C.
- We can use <u>disinfectants</u> on work surfaces and areas where germs thrive, such as toilets.
- <u>Antiseptics</u> can kill germs on living tissue, for example if we cut ourselves.
- <u>General good hygiene</u> is important in preventing the spread of disease. Examples include washing hands after going to the toilet, cooking food thoroughly and keeping food in the fridge or freezer. These methods prevent bacteria from multiplying and causing disease.

ARTIFICIAL IMMUNITY

- Artificial immunity involves the use of vaccines.
- <u>A vaccine contains dead or harmless germs</u>.
- These germs still have antigens on them and your white blood cells respond to them as if they were alive by multiplying and producing antibodies.
- A vaccine is an advanced warning so that if the person is infected by the germ the white blood cells can <u>respond immediately</u> and kill them.

DEFENCE AGAINST DISEASE

- The human body has many methods of <u>preventing pathogens</u> from entering the body.
- If pathogens do get into the body, however, your <u>immune system</u> goes into action.

NATURAL IMMUNITY

- Making antibodies takes time which is why you feel ill at first and then get better as the disease is destroyed by the white blood cells and antibodies.

- Once a particular antibody is made it stays in your body. If the same disease enters your body the antibodies are much quicker at destroying it and you feel no symptoms. <u>You are now immune to that disease</u>.

QUICK TEST

1. How does the skin protect against disease?
2. What is the job of mucus?
3. Name four ways of preventing the spread of germs.
4. Name the two types of white blood cell that are involved in the immune response.
5. How do phagocytes kill germs?
6. What chemicals do white blood cells produce?
7. What do antitoxins do?
8. What is the role of antibodies?
9. What are vaccines?
10. What are antibiotics and why are we always looking for new types?

1. It is a barrier and produces an antiseptic oil
2. It traps dust and germs
3. Sterilising, disinfectants, antiseptics and good hygiene
4. Phagocytes and lymphocytes
5. They engulf them
6. Antitoxins and antibodies
7. Destroy toxins
8. Attach to antigens and clump germs together for the phagocytes to engulf
9. Dead or weak forms of a disease
10. Antibiotics are used to treat bacterial infections; continual use make bacteria resistant, so new forms of antibiotic are needed

45

DRUGS, SOLVENTS, ALCOHOL AND TOBACCO

- **Smoking and solvents damage health, without a doubt.**
- **Alcohol and drugs are also dangerous if misused,**
  for either recreational and pharmaceutical purposes.

DRUGS – WHY ARE THEY DANGEROUS?

- Drugs are powerful chemicals; they alter the way the body works, often without you realising it.
- There are useful drugs such as penicillin and antibiotics, but these can be dangerous if misused.
- Some drugs affect the brain and nervous system, which in turn affect activities such as driving, behaviour and risk of infection.
- Drugs affect people in many different ways; you can never be sure what will happen to you.
- An overdose can easily happen by accident as it is difficult to tell how strong a drug is or how much to take.
- Drugs which affect the brain fall into four main groups:

SEDATIVES
- These drugs slow down the brain and make you feel sleepy. Tranquillisers and sleeping pills are examples.
- They are often given to people suffering from anxiety and stress.
- Barbiturates, which are powerful sedatives, are used as anaesthetics in hospitals.
- These drugs seriously alter reaction times and give you poor judgement of speed and distances.

PAINKILLERS
- These drugs suppress the pain sensors in the brain.
- Aspirin, heroin and morphine are examples.

- Morphine is given to people in cases of extreme pain.
- Heroin can be injected, which can increase the risk of contracting HIV; it is also highly addictive. People who become addicted to heroin often resort to crime to pay for the drug and suffer personality problems.

HALLUCINOGENS
- These drugs make you see or hear things that don't exist. These imaginings are called hallucinations.
- Examples are ecstasy, LSD and cannabis.
- The hallucinations can lead to fatal accidents.
- Ecstasy can give the user feelings of extreme energy. This extra energy can lead to a danger of overheating and dehydration.

STIMULANTS
- These drugs speed up the brain and nervous system and make you more alert and awake.
- Examples include amphetamines, cocaine and the less harmful caffeine in tea and coffee.
- Overuse results in high energy levels, changes in personality and hallucinations.
- Dependence on these drugs is high and withdrawing use causes serious depression.

Examiner's Top Tip
Concentrate on the health problems for the exam, but the social aspects are still important.

ALCOHOL

- Alcohol is a legal and socially acceptable drug, but it can still cause a lot of harm.
- Alcohol is a <u>depressant</u> and reduces the activity of the brain and nervous system.
- It is absorbed through the gut and taken to the brain in the blood.
- Alcohol damages neurones in the brain and causes irreversible brain damage.
- The liver breaks down alcohol at the rate of one unit an hour, but an excess of alcohol has a very <u>damaging</u> <u>effect</u> <u>on</u> <u>the</u> <u>liver</u>, <u>called</u> <u>cirrhosis</u>.
- Increasing amounts of alcohol cause people to lose control and slur their words. In this state accidents are more likely to happen.
- Alcohol can become very addictive without the person thinking they have a problem.

½ pint cider (0.3 litre)
1 glass of wine
1 glass sherry
1 single whisky
½ pint beer (0.3 litre)

All these drinks contain one unit of alcohol

SOLVENTS

- Solvents include everyday products like glue and aerosols.
- Solvent fumes are inhaled and are absorbed by the lungs. They soon reach the brain and <u>slow</u> <u>down</u> <u>breathing</u> <u>and</u> <u>heart</u> <u>rates</u>.
- Solvents also damage the <u>kidneys</u> <u>and</u> <u>liver</u>.
- Repeated inhalation can cause loss of control and unconsciousness.
- Many first-time inhalers die from heart failure or suffocation if using aerosols.
- Many of the symptoms are likened to being drunk, vomiting may occur and the person may not be in control.
- Solvents are also highly flammable.

ADDICTION

- Drugs, solvents, alcohol and tobacco can become <u>chemically</u> <u>or</u> <u>psychologically</u> <u>addictive</u>.
- In chemical addiction the body gets used to the drug (becomes <u>tolerant</u>) and the person has to take an increasing amount of the drug for it to continue to have an effect.
- If the person stops taking the drug then they develop <u>withdrawal</u> <u>symptoms</u> such as fever, nausea and hallucinations.
- In psychological addiction the person feels that they have to keep taking the drug but would not experience harmful effects if they stopped.

SMOKING

- Tobacco definitely causes health problems.
- It contains many harmful chemicals: <u>nicotine</u> is an addictive substance and a mild stimulant; <u>tar</u> is known to contain carcinogens that contribute to cancer; and <u>carbon</u> <u>monoxide</u> prevents the red blood cells from carrying oxygen.
- Some diseases aggravated by smoking include <u>emphysema</u>, <u>bronchitis</u>, <u>heart</u> <u>and</u> <u>blood</u> <u>vessel</u> <u>problems</u> <u>and</u> <u>lung</u> <u>cancer</u>.
- As well as health problems there is also the high cost of smoking and the negative social problems.

QUICK TEST

1. Which parts of the body are affected by alcohol?
2. What are stimulants?
3. Name three chemicals contained in tobacco.
4. What diseases does smoking aggravate?
5. What is the name of the disease of the liver?

1. Brain, liver and nervous system
2. Drugs that speed up the nervous system
3. Tar, nicotine and carbon monoxide
4. Emphysema, bronchitis, lung cancer and heart disease
5. Cirrhosis

DIABETES

Diabetes results when the <u>pancreas</u> <u>does</u> <u>not</u> <u>make</u> <u>enough</u> <u>of</u> <u>the</u> <u>hormone</u> <u>insulin</u>. As a consequence blood sugar levels rise and very little glucose is absorbed by the cells for respiration.

This can make the sufferer tired and thirsty.

If untreated, it leads to weight loss and even death.

<u>Diabetes</u> <u>can</u> <u>be</u> <u>controlled</u> <u>in</u> <u>two</u> <u>ways</u>:

- <u>Attention</u> <u>to</u> <u>diet</u>. A special low-glucose diet is needed and can be all that is needed to control some diabetes.
- In more severe cases, diabetics have to <u>inject</u> <u>themselves</u> <u>with</u> <u>insulin</u> before meals. This causes the liver to convert the glucose into glycogen straight away, thus removing glucose from the blood.

HORMONES AND DIABETES

- Hormones are <u>chemical</u> <u>messengers</u> produced by glands known as <u>endocrine</u> <u>glands</u>.
- Hormones <u>travel</u> <u>in</u> <u>the</u> <u>blood</u> to target organs.
- <u>Diabetes</u> is a disease caused by too little of the hormone insulin.

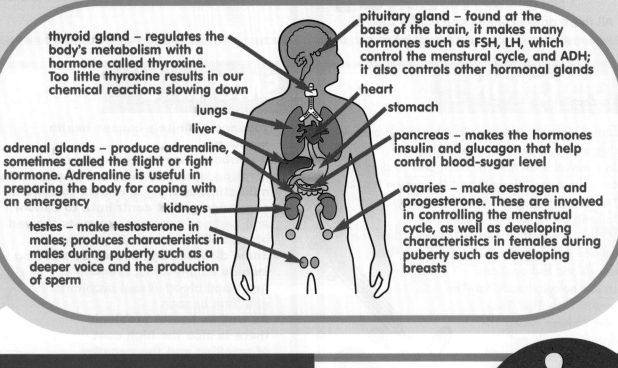

thyroid gland – regulates the body's metabolism with a hormone called thyroxine. Too little thyroxine results in our chemical reactions slowing down

pituitary gland – found at the base of the brain, it makes many hormones such as FSH, LH, which control the menstural cycle, and ADH; it also controls other hormonal glands

heart

lungs

stomach

liver

pancreas – makes the hormones insulin and glucagon that help control blood-sugar level

adrenal glands – produce adrenaline, sometimes called the flight or fight hormone. Adrenaline is useful in preparing the body for coping with an emergency

kidneys

ovaries – make oestrogen and progesterone. These are involved in controlling the menstrual cycle, as well as developing characteristics in females during puberty such as developing breasts

testes – make testosterone in males; produces characteristics in males during puberty such as a deeper voice and the production of sperm

NEGATIVE FEEDBACK

Negative feedback is involved in all homeostatic mechanisms. The pancreas controls blood sugar levels using this mechanism. Look at the diagram to see how it works.

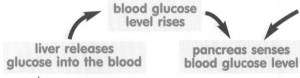

blood glucose level rises

blood sugar level drops

liver releases glucose into the blood

pancreas senses blood glucose level

liver takes up extra glucose from the blood

pancreas stops secreting insulin and secretes glucagon

pancreas secretes insulin

Examiner's Top Tip
It is common for people to confuse glycogen (storage glucose) with glucagon (the hormone). Make sure you know the difference.

PANCREAS AND HOMEOSTASIS

- Homeostasis is the mechanism by which the body <u>maintains</u> <u>normal</u> <u>levels</u> such as temperature and control of body water by making constant adjustments.
- The pancreas is an organ involved in homeostasis; it <u>maintains</u> <u>the</u> <u>level</u> <u>of</u> <u>glucose</u> <u>(sugar)</u> <u>in</u> <u>the</u> <u>blood</u> so that there is enough for respiration.
- The pancreas secretes two hormones into the blood, <u>insulin and glucagon</u>.
- If blood sugar levels are <u>too high</u>, which could be the case after a high carbohydrate meal, special cells in the pancreas detect these changes and release <u>insulin</u>.
- The <u>liver</u> responds to the amount of insulin in the blood and takes up glucose and <u>stores</u> <u>it</u> <u>as</u> <u>glycogen</u>. <u>Blood</u> <u>sugar</u> <u>levels</u> <u>return</u> <u>to</u> <u>normal</u>.

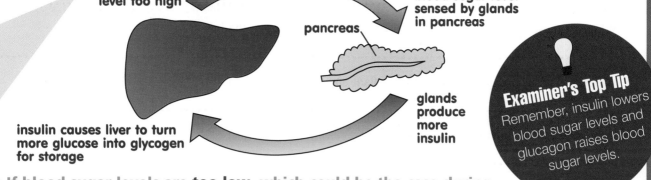

blood sugar level too high

pancreas

blood sugar level sensed by glands in pancreas

glands produce more insulin

insulin causes liver to turn more glucose into glycogen for storage

Examiner's Top Tip
Remember, insulin lowers blood sugar levels and glucagon raises blood sugar levels.

- If blood sugar levels are <u>too low</u>, which could be the case during exercise, the pancreas secretes <u>glucagon</u>.
- <u>Glucagon</u> stimulates the <u>conversion</u> <u>of</u> <u>stored</u> <u>glycogen</u> <u>in</u> <u>the</u> <u>liver</u> <u>back</u> <u>into</u> <u>glucose</u> which is released into the blood. <u>Blood</u> <u>sugar</u> <u>levels</u> <u>return</u> <u>to</u> <u>normal</u>.

HORMONE ACTION

- Hormones are just one way of transmitting information from place to place within the body.
- Hormonal effects tend to be <u>slower, long lasting</u> and <u>can affect a number of organs</u>, as is the case for adrenaline.
- Nervous control is <u>much quicker</u>, as in the reflex response, but its effects <u>don't last very long</u>. Nervous control is also <u>confined to a particular area</u>.

QUICK TEST

1. Which gland secretes adrenaline?
2. Which gland is involved in control of the menstrual cycle?
3. What is homeostasis?
4. What two hormones does the pancreas produce?
5. What other organ is involved in controlling blood sugar levels?
6. What does the liver do with excess glucose?
7. Which hormone raises blood sugar levels?
8. Which hormone lowers blood sugar levels?
9. What causes diabetes?
10. How can diabetes be treated?

1. Adrenal glands
2. Pituitary gland
3. Maintaining the body at normal levels, making constant adjustments
4. Insulin and glucagon
5. Liver
6. Convert it to glycogen for storage
7. Glucagon
8. Insulin
9. Pancreas not producing enough insulin
10. Attention to diet and injections of insulin

EXAM QUESTIONS - Use the questions to test your progress. Check your answers on page 94.

1. What do the letters that make up MRS GREN stand for?

....

2. Name the five main parts of a flowering plant.

....

3. What is the function of the stem?

....

4. What cells are contained inside the stem?

....

5. Which of these cells are responsible for transporting water up to the leaves?

....

6. Which cells transport glucose around the plant and where is it taken to?

....

7. What are the two functions of the roots?

....

8. What is photosynthesis?

....

9. When does photosynthesis take place?

....

10. When does respiration in a plant take place?

....

11. Write down the word equation for photosynthesis.

....

12. What three factors limit the rate of photosynthesis?

....

13. What is the role of the chloroplasts?

....

14. What adaptations does the root hair cell have for absorbing water from the soil?

....

15. Where on a plant would you find guard cells, and what is their function?

....

16. Label the diagram of a plant cell.

a............ c
............b
f............ ....d
............e

17. What is diffusion? Give an example of where it takes place in a plant.

....

18. What is a turgid cell?

....

19. What is a flaccid cell?

....

20. What is the transpiration stream?

....

21. Under what conditions is the transpiration stream fastest?

....

22. What three things are plants sensitive to in their surroundings?

....

23.What is auxin and where is it found?

....

24. Does auxin speed up or slow down the growth in roots?

....

25. What is this plant shoot responding to and how can you tell?

....

26. What are the advantages to commercial plant growers of synthetic hormones?

....

27. What are the plant's responses called?

....

28. Label the diagram of the leaf

a...........

b

c

d...........

e

g...........

f

29. A plant needs minerals such as nitrates, potassium and
phosphates for healthy growth.
   Match the plants to their deficiency symptoms:
a) Poor root growth and purple young leaves ....
b) Yellow leaves with dead spots ....
c) Stunted growth and yellow older leaves ....

1
lack of nitrates

2
lack of
phosphates

3
lack of
potassium

30. What other minerals does a plant need to make chlorophyll?

....

How did you do?

1–7	correct ....start again
8–15	correct ....getting there
16–22	correct ....good work
23–30	correct ....excellent

EXAM QUESTIONS - Use the questions to test your progress. Check your answers on page 94–95.

1. Label this animal cell:

b ....

c ....

.... a

2. Name three differences between a plant cell and an animal cell.

....

3. Name three similarities.

....

4. Give an example of where diffusion takes place in an animal cell.

....

5. What happens to animal cells if they are placed in pure water?

....

6. Why does this not happen to plant cells?

....

7. What is the process in which water moves from an area of high concentration to an area of low concentration?

....

8. What is active transport? And where does the energy come from?

....

9. What is the standard body temperature of humans?

....

10. Which parts of the body are affected by alcohol?

....

11. What diseases can be caused by smoking?

....

12. Name three microbes that can cause disease.

....

13. What is diabetes?

....

14. How can diabetes be treated?

....

15. The hormones glucagon and insulin are made in which organ of the body?

....

16. What is the function of:
a) Insulin? ....
b) Glucagon? ....

17. What is homeostasis?

....

18. What happens to your blood vessels when your body is hot?

....

19. What other responses take place in hot conditions?

....

20. What part of our brain controls the body's reaction to temperature changes?

....

21. What is hypothermia, and who is most susceptible?

....

22. Label this picture of a virus

a............

b............

23. What is a vector? Give an example.

....

24. What role does the liver play in controlling blood sugar levels?

....

25. What is urea?

....

....

26. How is it removed from the body?

....

....

27. What important roles do the kidneys have?

....

....

28. What are the two types of white blood cells that protect your body against disease?

....

....

29. What is natural immunity?

....

....

30. What is artificial immunity?

....

....

How did you do?

1–7	correct	....start again
8–15	correct	....getting there
16–22	correct	....good work
23–30	correct	....excellent

THE MENSTRUAL CYCLE

The menstrual cycle lasts approximately <u>28 days</u>. It consists of a <u>menstrual bleed</u> <u>and</u> <u>ovulation</u> – the release of an egg. <u>Hormones</u> control the whole cycle. Ovaries secrete the hormones <u>progesterone</u> and <u>oestrogen</u>.

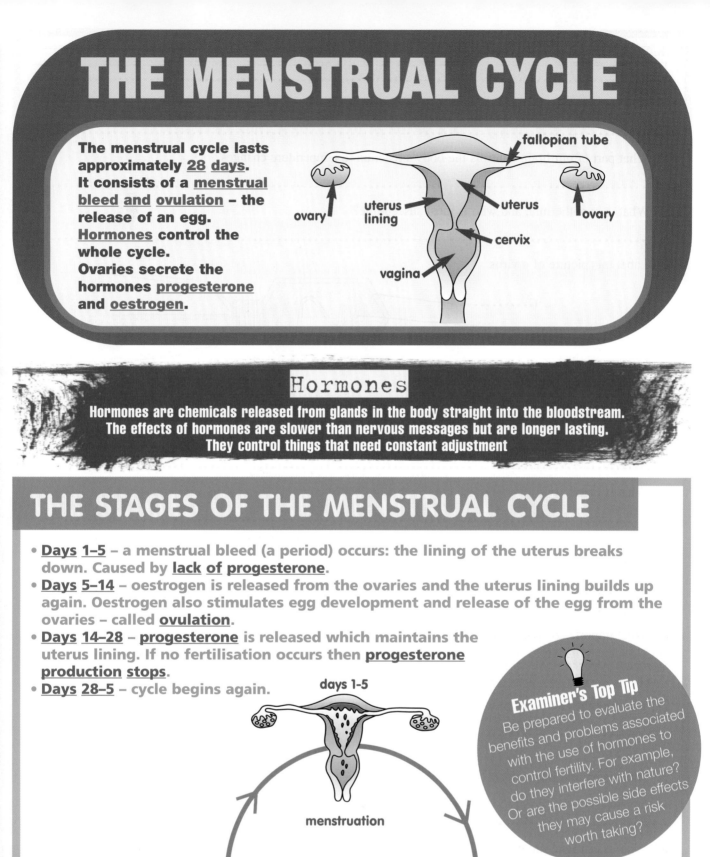

Hormones

Hormones are chemicals released from glands in the body straight into the bloodstream. The effects of hormones are slower than nervous messages but are longer lasting. They control things that need constant adjustment

THE STAGES OF THE MENSTRUAL CYCLE

- <u>Days</u> <u>1–5</u> – a menstrual bleed (a period) occurs: the lining of the uterus breaks down. Caused by <u>lack</u> <u>of</u> <u>progesterone</u>.
- <u>Days</u> <u>5–14</u> – oestrogen is released from the ovaries and the uterus lining builds up again. Oestrogen also stimulates egg development and release of the egg from the ovaries – called <u>ovulation</u>.
- <u>Days</u> <u>14–28</u> – <u>progesterone</u> is released which maintains the uterus lining. If no fertilisation occurs then <u>progesterone</u> <u>production</u> <u>stops</u>.
- <u>Days</u> <u>28–5</u> – cycle begins again.

Examiner's Top Tip
Be prepared to evaluate the benefits and problems associated with the use of hormones to control fertility. For example, do they interfere with nature? Or are the possible side effects they may cause a risk worth taking?

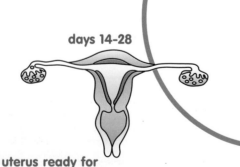

days 1-5

menstruation

days 14-28

days 5-14

uterus ready for implantation

ovulation

THE PITUITARY GLAND

- The two hormones released from the ovaries are controlled by the <u>pituitary gland</u> situated at the base of the brain.
- The pituitary gland secretes two more hormones, follicle stimulating hormone and luteinising hormone. There is no need to learn how to spell them as they can be written as FSH and LH.

Follow the diagram below to see how the hormones interact to control the menstrual cycle.

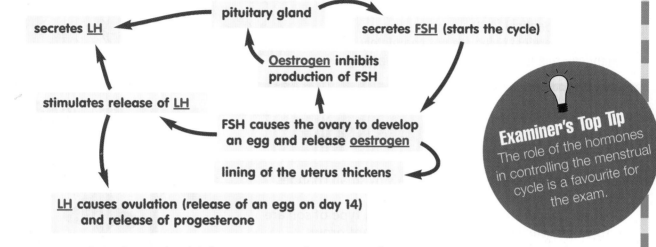

Examiner's Top Tip
The role of the hormones in controlling the menstrual cycle is a favourite for the exam.

<u>Progesterone</u> is released, which maintains the uterus lining.
<u>Oestrogen</u> also keeps the uterus lining thick, ready for pregnancy.
If no pregnancy occurs, progesterone production stops and the cycle begins again.

CONTROLLING FERTILITY

Fertility in women can be controlled in two ways:
- *FSH can be administered as a '<u>fertility drug</u>' to women whose own production is too low to stimulate eggs to mature. This can result in multiple births.*
- *Oestrogen can be used as an <u>oral contraceptive</u> to inhibit FSH so that no eggs mature.*

QUICK TEST

1. On what days does the menstrual bleed usually take place?
2. What causes the uterus lining to break down?
3. Where are the hormones oestrogen and progesterone made?
4. From where are the eggs released?
5. Although oestrogen is stated as stimulating egg release, this is actually caused by another hormone. Which one?
6. Which two hormones are produced by the pituitary gland?
7. What two things does follicle stimulating hormone do?
8. Which two hormones maintain the uterus lining?
9. What is ovulation?
10. How long is the average menstrual cycle?

10. 28 days
9. Release of an egg
8. Progesterone and oestrogen
7. Development of an egg and release of oestrogen
6. Luteinising hormone/LH and follicle stimulating hormone/FSH
5. Luteinising hormone/LH
4. Ovaries
3. Ovaries
2. Lack of progesterone
1. Days 1–5

GENETIC VARIATION

- Why do we look like we do? The answer is because we have inherited our characteristics from our parents.
- Brothers and sisters are not exactly the same as each other because they inherit different genes from their parents. It is completely random.
- There are thousands of different genes in every human cell, so the combination of genes in a cell is endless; the chance of two people having the same genes is virtually impossible.
- Identical twins are an exception as their genes are identical.
- But even identical twins are not completely identical; this is due to environmental factors.

VARIATION IN PLANTS

Plants are affected more than animals by small changes in the environment.
Sunlight, temperature, moisture level and type of soil are factors that will determine how well a plant grows.
A plant grown in sunlight will grow much faster and may double in size compared to a plant grown in the shade, whereas a dog living in England would show no significant changes if it moved to Africa.

VARIATION IN ANIMALS

- We vary because of the random way our genes are inherited.
- The environment can affect most of our characteristics. It is usually a combination of genetics and environment that determines how we look and behave.
- Just how significant the environment is in determining our features is difficult to assess; for example, is being good at sport inherited or is it due to your upbringing?
- There are some characteristics that are not affected by the environment at all:
1. Eye colour   2. Natural hair colour   3. Blood group   4. Inherited diseases

HOW CAN WE TELL IF THE DIFFERENCES ARE DUE TO THE ENVIRONMENT?

We can **produce** **clones** of plants by taking cuttings. The cuttings are genetically identical to each other.
We can then grow the plants in different conditions.
Any differences in their appearance would be due to the environment.
If you couldn't produce clones and wanted to test whether the differences were due to environmental or genetic reasons, then you could:
- Replant them in a different place to see if they then grew more alike. If they did it was the environment that caused the difference.
- Plant two plants of the same species in the same conditions to see if the environment caused the differences again.

ENVIRONMENTAL VARIATION

- The environment is your surroundings and all the things that may affect your upbringing.
- Identical twins may be separated at birth and grow up in totally different surroundings, following different diets for example.
- Any differences between the twins must be due to the environment they were brought up in as they have identical genes.
- Many of the differences between people are caused by a combination of genetic and environmental influences.

VaRIATioN

- All living things vary in the way they look or behave.
- Variation can be between species or within species.
- Living things that belong to the same species are all slightly different.
- Genetics, the environment or a combination of both may cause these differences.

CONTINUOUS AND DISCONTINUOUS VARIATION

- Differences between animals and plants show two types of variation.
- If you measured the heights of people in your class you would find that they varied gradually from short to tall.
- Height or weight follows continuous variation.

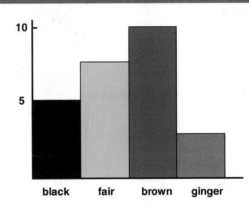

black   fair   brown   ginger

- If you looked at the hair colour of people in your class, you would find there are only a few options, not a continuous range.
- Another example is whether a person can roll their tongue. You either can or you cannot; there is no in between.
- Eye colour, hair colour, blood group and rolling tongues are examples of discontinuous variation.

- - - - - - -

QUICK TEST

1. Is blood group inherited or caused by the environment?
2. Is having a scar environmental or inherited?
3. Why do animals and plants of the same species vary?
4. Give two examples of continuous variation.
5. Give two examples of discontinuous variation.
6. What four environmental factors determine the growth of plants?

1. Inherited
2. Environmental
3. Random inheritance of parental genes and environmental effects
4. Height and weight
5. Eye colour, blood group, hair colour, tongue rolling (any two)
6. Amount of sunlight and moisture, temperature and type of soil

GENETICS

Genetics is the study of how information is passed on from generation to generation. Genetic diagrams are used to show how certain characteristics are passed on.

MENDEL'S EXPERIMENTS

- Gregor Mendel, an Austrian monk, discovered the principle behind genetics by studying the inheritance of a single factor in pea plants.
- The inheritance of single characteristics is called <u>monohybrid</u> <u>inheritance</u>.

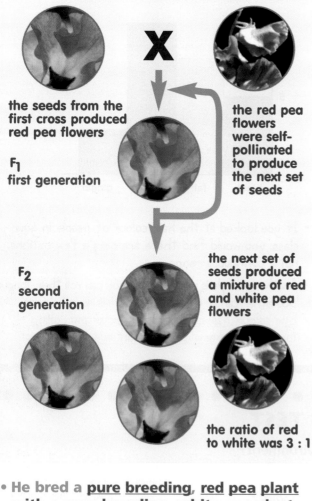

the cross shows that plants were bred together

X

the seeds from the first cross produced red pea flowers

F_1 first generation

the red pea flowers were self-pollinated to produce the next set of seeds

F_2 second generation

the next set of seeds produced a mixture of red and white pea flowers

the ratio of red to white was 3 : 1

- He bred a <u>pure</u> <u>breeding</u>, <u>red pea plant</u> <u>with a pure</u> <u>breeding</u>, <u>white pea plant</u> and found that they always produced red flowers (the <u>F_1 generation</u>).
- He named the red feature the <u>dominant</u> <u>characteristic</u>.

- When he bred two of the red pea plants together, he discovered that the next set of flowers were a mixture of red and white pea flowers (<u>the F_2 generation</u>).
- The <u>ratio</u> of red to white was <u>3 : 1</u>.
- Mendel called the white characteristic <u>recessive</u>.
- From his experiments Mendel concluded that the peas must carry a <u>pair of</u> <u>factors</u> <u>for</u> <u>each</u> <u>feature</u>.
- When the seeds were formed, they inherited one factor from each parent at random.
- We now call these factors <u>genes</u>. Genes occur on <u>pairs</u> <u>of</u> <u>chromosomes</u>. Each form of a gene is called an <u>allele</u>.
- We can show the results of Mendel's pea plant cross using symbols.
- The <u>dominant</u> characteristic is given a <u>capital</u> <u>letter</u> and the <u>recessive</u> characteristic a <u>lower</u> <u>case</u> <u>letter</u>.
- In this example, the letter 'R' represents red flowers and 'r' represents white flowers.

Mendel's crosses in symbols

parents (two factors for each characteristic)	RR red	X	rr white
gametes (one factor for each pair)	R or R		r or r
first generation (F_1)		Rr both red	
parents	Rr		Rr
gametes	R or r		R or r
second generation (F_2)	rR RR rr Rr		3 red 1 white

A WORKED EXAMPLE – INHERITANCE OF EYE COLOUR

Remember, we have two alleles for eye colour, one from each parent, making up a gene.
- The allele for brown eyes is <u>dominant</u>, so it can be represented by the letter 'B'.
- The allele for blue eyes is <u>recessive</u> so it is represented by the letter 'b'.
- If the mother and father are both <u>heterozygous</u> for eye colour, they have the genotype 'Bb'.

What colour eyes will their children have?
- We can show the possible outcomes using a <u>Punnett</u> <u>square</u>.

<u>Different</u> <u>combinations</u> of <u>genotypes</u> can be crossed and the outcomes worked out in the same way using a <u>Punnett</u> <u>square</u>.
- For example, what would happen if the mother was <u>heterozygous</u> for eye colour (Bb) and the father was <u>homozygous</u> for blue eyes (bb)?
- This gives a 1 : 1 ratio of brown eyes to blue eyes.

<u>Remember, the ratios are only probabilities and may not always happen</u>.

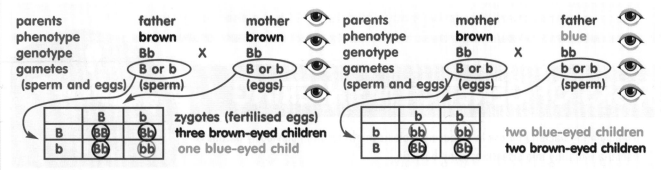

parents	father	mother
phenotype	**brown**	**brown**
genotype	**Bb** X	**Bb**
gametes (sperm and eggs)	**B or b** (sperm)	**B or b** (eggs)

	B	b
B	BB	Bb
b	Bb	bb

zygotes (fertilised eggs)
three brown-eyed children
one blue-eyed child

parents	mother	father
phenotype	**brown**	blue
genotype	**Bb** X	**bb**
gametes (sperm and eggs)	**B or b** (eggs)	**b or b** (sperm)

	b	b
b	bb	bb
B	Bb	Bb

two blue-eyed children
two brown-eyed children

- This gives a 3 : 1 ratio of brown to blue eyes.

DEFINITIONS

- <u>Recessive</u> means it is the weaker allele and only has an effect in the homozygous recessive condition.
- <u>Dominant</u> means it is the stronger allele and has an effect in the heterozygous condition.
- The <u>genotype</u> is the type of alleles an organism carries; the genotype of the red pea plants could be RR or Rr. Although the genotypes are different they are still red because red is dominant.
- The <u>phenotype</u> is what the plant physically looks like, the result of what genotype the organism has.
- If the pea plant has both alleles the same, it is <u>homozygous</u> <u>dominant</u> (RR) or <u>homozygous</u> <u>recessive</u> (rr).
- If the pea plant has different alleles, it is <u>heterozygous</u> (Rr).

QUICK TEST

1. A plant with a genotype RR is homozygous/heterozygous dominant?

2. A person with blue eyes has a genotype/phenotype bb?

3. A plant that has two different alleles is heterozygous or homozygous?

4. What does recessive mean?

5. What does dominant mean?

6. What is a zygote?

7. What are gametes?

8. What does monohybrid inheritance mean?

9. Predict the outcome of a cross between two blue-eyed homozygous recessive parents.

10. What would be the genotype of a brown-eyed child if brown was the dominant characteristic? (B = brown, b = blue)

10. BB or Bb
9. All offspring/children will have blue eyes
8. The inheritance of one characteristic
7. Sperms or eggs
6. A fertilised egg
5. The stronger allele
4. The weaker allele
3. Heterozygous
2. Genotype
1. Homozygous

GENETIC ENGINEERING

GENE THERAPY

- It may be possible to use genetic engineering to treat inherited diseases such as **cystic fibrosis**.
- Sufferers could be cured if the correct gene could be inserted into their body cells.
- The problem that exists with cystic fibrosis is that the cells that need the correct gene are in many parts of the body, which makes it difficult to remove them to insert the required gene.
- Another problem is that even if the correct gene was inserted into the body cells, the cells would not multiply.
- This means that there would be many cells that still have the faulty gene.

RISKS OF GENETIC ENGINEERING

- Manipulating bacteria for use in producing proteins might result in previously harmless bacteria mutating into disease-causing bacteria.
- Gene therapy is potentially a way forward in curing fatal diseases, but it poses risks as inserting genes into human cells may make them cancerous.
- There is a possibility that a human egg can be taken out of the womb and the harmful genes removed, before it is inserted back to continue its growth into a human.
- Genetic engineering is seen by many as manipulating the 'stuff of life'; is it morally and ethically wrong to interfere with nature?
- There is still a lot of unease about methods used as nobody can be completely sure what the results will be.

DNA AND PROTEIN SYNTHESIS

- A gene is a length of DNA containing a small part of the genetic code.
- Genes control which protein a cell makes and therefore control the development, structure and function of the whole organism.
- Proteins made up of amino acids control our characteristics.
- Just how genes code for a particular protein is explained in a process called protein synthesis.
- Protein synthesis begins in the nucleus and ends in the cytoplasm of cells.
- You can see the structure of DNA in more detail in the topic on Genes, Chromosomes and Mutations.

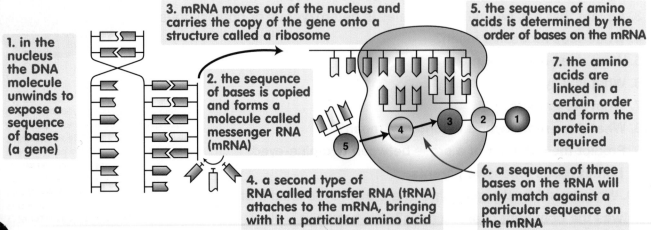

1. in the nucleus the DNA molecule unwinds to expose a sequence of bases (a gene)

2. the sequence of bases is copied and forms a molecule called messenger RNA (mRNA)

3. mRNA moves out of the nucleus and carries the copy of the gene onto a structure called a ribosome

4. a second type of RNA called transfer RNA (tRNA) attaches to the mRNA, bringing with it a particular amino acid

5. the sequence of amino acids is determined by the order of bases on the mRNA

6. a sequence of three bases on the tRNA will only match against a particular sequence on the mRNA

7. the amino acids are linked in a certain order and form the protein required

- Genetic engineering is the process in which genes from one organism are removed and inserted into the cells of another.
- It has many exciting possibilities, but is not without its problems.
- Scientists can now genetically modify plants and animals using the process of genetic engineering.

MANIPULATING GENES

- You have seen how our genes code for a particular protein that enables all our normal life processes to function.
- Many diseases are caused when the body cannot make a particular protein.
- Genetic engineering has been used to treat diabetic people through the production of the protein, insulin.
- The gene that codes for insulin can be found in human pancreas cells.
- Other human proteins made in this way include the human-growth hormone that is used to treat children who do not grow properly.

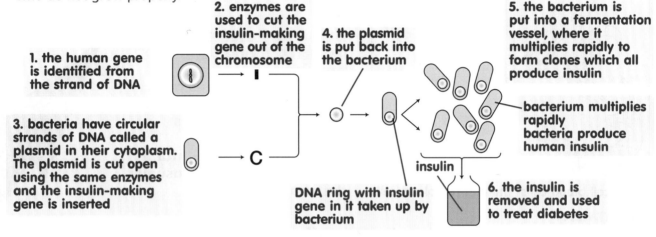

1. the human gene is identified from the strand of DNA

2. enzymes are used to cut the insulin-making gene out of the chromosome

3. bacteria have circular strands of DNA called a plasmid in their cytoplasm. The plasmid is cut open using the same enzymes and the insulin-making gene is inserted

4. the plasmid is put back into the bacterium

DNA ring with insulin gene in it taken up by bacterium

5. the bacterium is put into a fermentation vessel, where it multiplies rapidly to form clones which all produce insulin

bacterium multiplies rapidly
bacteria produce human insulin

insulin

6. the insulin is removed and used to treat diabetes

THE BENEFITS OF GENETIC ENGINEERING

- Genetic engineering benefits industry, medicine and agriculture in many ways.
- We have developed plants that are resistant to pests and diseases and plants that can grow in adverse environmental conditions.
- Wheat and other crops have been developed that can take nitrogen from the air directly and produce proteins without the need for costly fertilisers.
- Tomatoes and other sorts of fruit are now able to stay fresher for longer.
- Animals are engineered to produce chemicals in their milk, such as drugs and human antibodies.
- The list seems endless and there are no doubts as to the benefits of genetic engineering now and in the future, but there are also risks and moral issues that are associated with this relatively modern technology.

Tomatoes can be genetically engineered to stay fresh for longer by inserting a gene from fish into their cells. Would you eat one?

Examiner's Top Tip
Protein synthesis will only appear in the higher paper as it is a difficult concept.

QUICK TEST

1. What do genes code for?

2. What are genes made of?

3. Give an example of the use of bacteria to treat a disease.

4. What are the risks associated with using bacteria to produce human proteins?

5. What is gene therapy?

5. Using genetic engineering to treat inherited diseases
4. The bacteria may mutate into harmful bacteria
3. Making human insulin
2. DNA
1. A particular characteristic

SICKLE CELL ANAEMIA

CAUSES

Sickle cell anaemia is caused by a <u>recessive</u> <u>allele(s)</u>.
Two carrier parents have a one in four chance
of having a child with the disease.
The child will be <u>homozygous</u> <u>recessive (ss)</u>.

SYMPTOMS

• Sickle cell is an inherited disease of the <u>blood</u>.
• The red blood cells are an <u>abnormal</u> <u>shape</u>,
  called sickle shape.

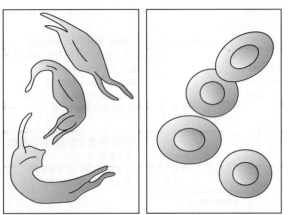

sickle cell blood cells   normal red blood cells

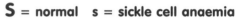

S = normal   s = sickle cell anaemia

	S	s
s	Ss	ss
S	SS	Ss

• **<u>The sickle shape affects the
  oxygen carrying capacity of
  the blood</u>.**
• The sickle blood cells get
  stuck in the capillaries, which
  deprives the body cells
  of oxygen.
• It is an extremely painful
  disease and sufferers usually
  die at an early age.

TREATMENT

• There is no cure and even though many sufferers die before they can reproduce,
  the disease does not disappear, especially in Africa.
• This is because <u>heterozygous</u> <u>carriers (Ss)</u> are immune to malaria.
• The microbe that causes malaria does not affect sickle red blood cells in the same
  way as normal red blood cells, so carriers do not die from maleria.
• <u>Being a carrier is advantageous in malarial regions</u>.

HUNTINGDON'S CHOREA

CAUSES

• Huntington's chorea is caused by a <u>dominant allele (H)</u>.
• This means only one allele is needed to pass on the disease,
  so all <u>heterozygous people are sufferers (Hh)</u>.
• The only people free from the disease are <u>homozygous recessive (hh)</u>.
• It affects one in 20000 people, so is a rare disease.
• There is a 50% chance of inheriting the disease if just one
  parent is a carrier.

**50% chance of
having the disease**

SYMPTOMS

<u>The brain degenerates and the sufferer has uncontrolled, jerky</u> movements.
The sufferer becomes moody and depressed and their memory is affected.

	h	h
h	hh	hh
H	Hh	Hh

H = Huntington's chorea
h = normal

TREATMENT

There is no cure.
Onset of the disease is late; the sufferer is about 30–40 years old before they realise that
they have it. Consequently many have already had children and passed on the disease.

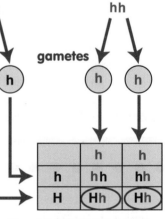

CYSTIC FIBROSIS

CAUSES

- Cystic fibrosis is the commonest inherited disease in Britain; about one in 2000 children born in Britain has cystic fibrosis.
- Cystic fibrosis is caused by a <u>recessive allele (c)</u>, carried by about one person in 20. This makes the chances of two carriers marrying one in 400.
- People who are <u>heterozygous</u> with the genotype (Cc) are said to be <u>carriers</u>. They have no ill effects.
- Only people who are <u>homozygous</u> for this allele (cc) are affected.

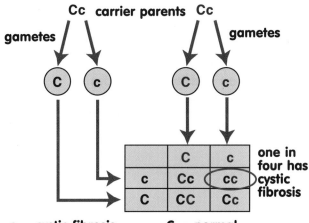

c = cystic fibrosis   C = normal

SYMPTOMS

- Cystic fibrosis sufferers produce large amounts of <u>thick</u>, <u>sticky</u> <u>mucus</u> that can block air passages and digestive tubes.
- The child has difficulty breathing and absorbing food.
- The mucus slows down the exchange of oxygen and carbon dioxide between the lungs and blood.
- The mucus also <u>encourages</u> <u>bacteria to grow</u>, which cause chest infections.

TREATMENT

- There is still no cure; treatment involves <u>physiotherapy</u> to try to remove some of the mucus and <u>strong</u> <u>antibiotics</u> to treat the infections.
- It was only in 1989 that the gene that causes cystic fibrosis was discovered.
- This allowed a test to be developed, which can tell if a person is a carrier by analysing their DNA.
- If a couple know that they are carriers they have a difficult decision to make. A <u>genetic counsellor</u> can explain to them that there is a one in four chance of having a child with cystic fibrosis. They then have to decide if the risk of having children is too great.

INHERITED DISEASES

Not all diseases are caused by pathogens. Genes pass on characteristics from one generation to the next. Sometimes 'faulty' genes are inherited that cause diseases. Inherited diseases are called <u>genetic diseases</u>.

QUICK TEST

1. How is cystic fibrosis inherited?

2. What are the symptoms of cystic fibrosis?

3. How is cystic fibrosis treated?

4. What are the chances of two carrier parents. having a child who suffers from cystic fibrosis?

5. What is sickle cell anaemia?

6. What are the symptoms of the disease?

7. Why is it advantageous to be heterozygous for sickle cell anaemia?

8. What are the symptoms of Huntington's chorea?

9. Why are the chances of inheriting Huntington's chorea so high?

10. Are there any cures for these diseases? What can be done instead?

10. No. Treatment of symptoms and genetic counsellor's advice
9. Caused by a dominant gene and sufferers have children before they realise they have it
8. Uncontrolled, jerky movements, depression,
7. Immunity against malaria
6. Lack of oxygen to the muscles and severe pain
5. A disorder of the red blood cells
4. One in four
3. By physiotherapy and strong antibiotics
2. Thick, sticky mucus, difficulty breathing, frequent infections
1. By a recessive allele passed on from two carrier parents

MITOSIS

The chromosomes at the beginning of mitosis look like a lot of tangled threads, but as the cell begins to divide they become visible. The cells in humans begin with 46 chromosomes.

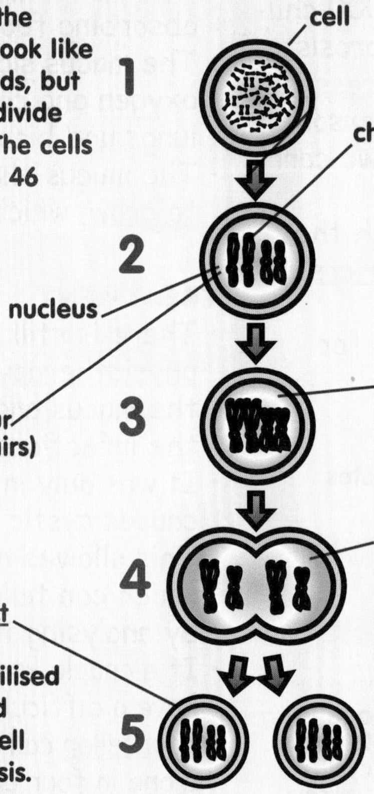

1

cell

chromosome

2

nucleus

for simplicity, only four chromosomes (two pairs) are shown in this cell

3

the chromosomes replicate themselves and for a while there are 92 chromosomes in the nucleus

4

the chromosomes pull apart and the cell divides into two cells, each with 46 chromosomes; these become the daughter cells

The daughter cells are exact copies of the original cell. If the parent cell was a fertilised egg, a new individual would develop as a result of the cell repeatedly dividing by mitosis.

5

DNA REPLICATION

· Just before a cell begins to divide the chromosomes have to be duplicated.
· The chromosomes are made up of long strands of deoxyribonucleic acid – DNA for short!
· DNA has the ability to copy itself exactly.

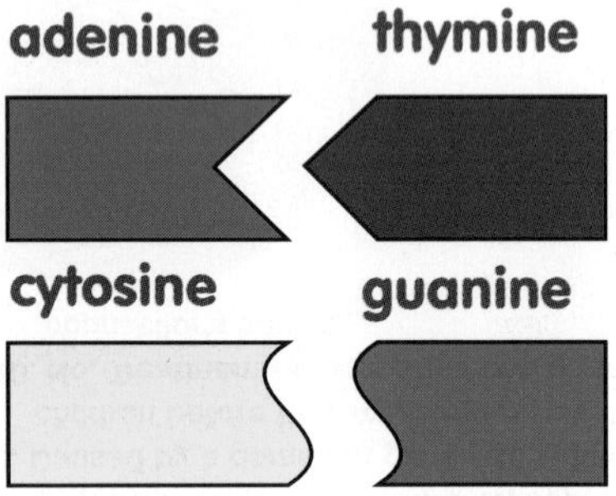

adenine   thymine

cytosine   guanine

(often just called A, T, C and G)

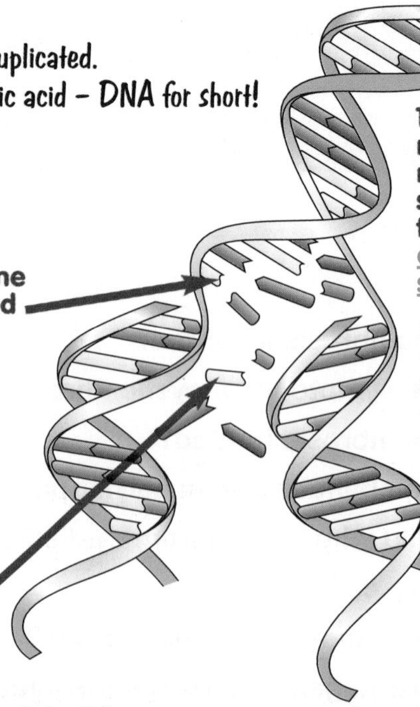

3. this leaves the exposed bases adenine (A), thymine (T), guanine (G) and cytosine (C)

1. the DNA molecule is made up of two strands coiled together in a double helix shape

2. the helix unwinds and the two strands are unzipped in the middle

4. new bases present in the cytoplasm join up with the exposed bases; A always pairs with T and G always pairs with C

5. in this way two identical strands of DNA are formed

6. the DNA coils back up into a double helix and the chromosomes have been copied

- • How do we grow from a fertilised egg?
- • How do we replace cells when we have cut ourselves?
- • The answer is cell division, called <u>mitosis</u>.
- • Mitosis produces all cells except the sex cells.
- • <u>Mitosis occurs in growth and replacement of cells</u>.

MITOSIS

Mitosis is when a cell reproduces to produce <u>two daughter cells that are identical to the original parent cell</u>.

MITOSIS AND ASEXUAL REPRODUCTION

- • *There are two types of reproduction, <u>sexual reproduction</u> and <u>asexual reproduction</u>.*
- • <u>*Asexual reproduction*</u> *involves only <u>one parent</u>.*
- • *The offspring have exact copies of the parents' genes and are called clones.*
- • *Asexual reproduction uses mitosis to produce clones.*
- • *Amoeba, a single-celled organism, reproduces asexually.*
- • *Some plants can reproduce asexually, e.g. strawberries, potatoes and daffodils.*
- • <u>*Sexual reproduction involves fertilisation*</u> *and produces offspring that are not genetically identical to the parents.*
- • *Sexual reproduction always involves a male and a female gamete produced by meiosis.*

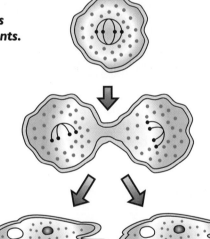

amoeba

becomes round in shape

and divides in two

QUICK TEST

1. How many daughter cells does mitosis produce?
2. How many chromosomes do the daughter cells in humans have?
3. What is mitosis used for?
4. What are chromosomes made of?
5. Does asexual reproduction need one parent or two?
6. Does asexual reproduction produce variation in individuals?
7. What type of individuals does it produce?
8. Give examples of organisms that reproduce using asexual reproduction.

8. Amoeba, daffodils, potatoes and strawberries
7. Clones
6. No
5. One
4. Coiled strands of DNA
3. Growth and replacement of cells
2. 46
1. Two

MEIOSIS AND FERTILISATION

there are 23 pairs of chromosomes at the start of meiosis, so 46 in total in the reproductive cells in the ovary or testis

cell

for simplicity, only four chromosomes (two pairs) are shown in this cell

a diploid cell has 46 chromosomes

nucleus

chromosome

one of each pair is from the father (shown yellow) and the other from the mother (shown pink)

the chromosomes make an exact copy of themselves

the chromosomes split up into two cells; this is done randomly so each cell has a mixture of the father's and mother's chromosomes

the chromosomes then pull apart to form four cells each, with half the number of chromosomes as the original cell (23)

gametes (sperm or egg cells) are produced

the gametes produced are called haploid gametes and have 23 chromosomes

Examiner's Top Tip
Learn what happens to the chromosomes in each stage of meiosis; remember the chromosome pairs are separated first and then the chromosomes themselves.

- Meiosis is a type of cell division that occurs in the formation of gametes (sex cells).
- Meiosis produces cells that have half the number of chromosomes to the original.
- Cells that have half the number of chromosomes (23 in humans) are called haploid cells.
- Fertilisation restores the normal number of chromosomes (46, diploid).

FERTILISATION

- In fertilisation a male gamete joins with a female gamete to produce a fertilised egg cell called a zygote.
- During fertilisation the 23 single chromosomes in the sperm cell pair up with the 23 chromosomes in the egg cell. They pair up with the opposite numbers, i.e. number four with number four.
- Fertilisation restores the number of chromosomes to the diploid number of 46, or 23 pairs.
- During meiosis it is a matter of chance which chromosomes make up the sperm and the egg and during fertilisation it is also a matter of chance which sperm fertilises which egg.
- Meiosis and fertilisation give rise to variation in the individual, as they will inherit a combination of the father's and mother's genes.

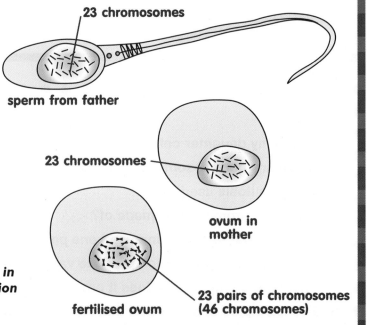

23 chromosomes

sperm from father

23 chromosomes

ovum in mother

fertilised ovum

23 pairs of chromosomes (46 chromosomes)

THE INHERITANCE OF SEX

- There are 23 pairs of human chromosomes.
- The <u>23rd</u> <u>pair</u> <u>are</u> <u>the</u> <u>sex</u> <u>chromosomes</u>; they determine whether you are a boy or a girl. All the other chromosomes contain information for your characteristics.
- If you are a male, one of the sex chromosomes is shorter than the other: this is the Y chromosome.
- Females have two X chromosomes that are the same size.

X	Y

male sex chromosome

X	X

female sex chromosome

- The female ovary will produce only X chromosomes during meiosis.
- The male testis will produce half X chromosome sperms and half Y chromosome sperms.
- During fertilisation the egg may join with either the X sperm or the Y sperm.

We can show this in a Punnett square diagram:

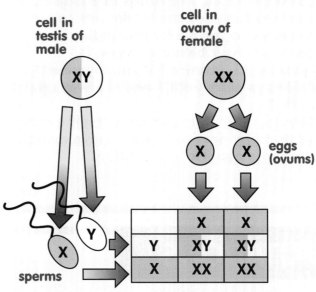

- The diagram shows that each time a couple have a child, there is a 50% chance that it will be male and a 50% chance that it will be female.
- The 50% chance of a boy or a girl is only a probability, all of the children could be girls or all boys.

Examiner's Top Tip
The inheritance of sex is a common exam question, so make sure you learn the diagram.

QUICK TEST

1. How many chromosomes are there in a haploid body cell?
2. How many chromosomes are there in a diploid cell?
3. What are gametes?
4. What is a zygote?
5. What is the probability of a couple's first child being a boy?
6. What is the probability of the second child being a girl?
7. If the 23rd pair of chromosomes are XY, is the child male or female?
8. How many chromosomes would an egg cell of a cat have if the body cell of a cat has 38 chromosomes?

8. 19
7. Male
6. 50%
5. 50%
4. A fertilised egg
3. The sex cells, sperms or eggs
2. 46
1. 23

DNA MOLECULE

- A coiled up DNA molecule makes up the 'arms' of a chromosome.
- A DNA molecule is joined together by chemical bases, like rungs in a ladder.
- The two rungs of the ladder are coiled together to form a **double helix**.
- There are four bases: adenine (A), cytosine (C), guanine (G) and thymine (T).
- A always pairs with T and G always pairs with C.
- **DNA has the ability to copy itself exactly, so that any new cells made have exactly the same genetic information.**

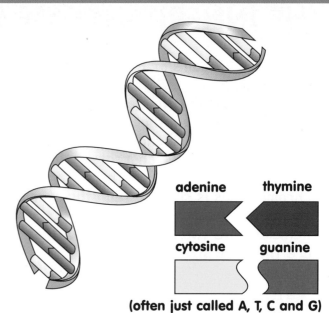

adenine   thymine

cytosine   guanine

(often just called A, T, C and G)

MUTATIONS

DNA is a very stable molecule, but occasionally things can go wrong in the copying process.

A mutation is a change in the chemical structure of a gene or chromosome, which alters the way an organism develops.

The change may happen for no reason or there might be a definite cause.

Mutations occur naturally in the environment; for example, new strains of the flu virus are always appearing. The chances of mutation can be increased by:

- Ionising radiation, such as alpha, beta and gamma radiation, may damage the DNA inside cells.
- X-rays and ultraviolet light may alter the genes in such a way that when they divide to replace themselves, uncontrollable growth is produced which develops into tumours and cancer.
- Certain chemicals called mutagens. Cancer-causing chemicals are called carcinogens and are found in cigarette smoke, for example.

radiation

INHERITING MUTATIONS

Mutations that occur in body cells are not inherited; they are only harmful to the person whose body cells are altered.

Mutations that occur in reproductive cells are inherited; the child will develop abnormally or die at an early age.

Some mutations that are inherited are beneficial and form the basis of evolution.

Organisms that become adapted to the environment are better able to survive and pass on their genes.

The peppered moth's allele for dark colouring gave it an advantage over the white moth when its environment became more polluted.

DOWN'S SYNDROME

- A chromosome mutation is a change in the number of chromosomes in the cell.
- When cells divide to form gametes (sex cells), they share out equally the number of chromosomes, so each has 23.
- Occasionally an extra chromosome goes into one cell and the fertilised egg will have 47 chromosomes instead of 46.
- The child has three chromosome number 21s instead of two in their cells and will suffer from Down's syndrome.
- The child can live a relatively normal life, but will be mentally retarded and more susceptible to some diseases.

GENES, CHROMOSOMES AND MUTATIONS

each nucleus contains thread like chromosomes

the chromosomes occur in pairs, one from the mother and one from the father

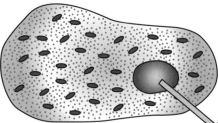

each chromosome is made up of a long stranded molecule called DNA

a gene is a section of DNA

there is a pair of genes for each feature. We call the different versions of a gene alleles

proteins and enzymes control all our characteristics; genes are chemical instructions that code for a particular protein or enzyme and therefore our characteristics

Examiner's Top Tip
It is crucial to understand where the chromosomes and genes originate from in order to understand the rest of the inheritance topics.

- Inside nearly all cells is a <u>nucleus</u>.
- The nucleus <u>contains</u> <u>instructions</u> that control all your characteristics.
- The instructions are carried on <u>chromosomes</u>.
- Genes on the chromosome control each particular characteristic.
- Inside human cells there are <u>23 pairs</u> or <u>46 chromosomes</u>. The cell is called <u>a diploid cell</u>.
- Other animals have different numbers of chromosomes.

QUICK TEST

1. What are the thread-like features contained in the nucleus called?

2. How many chromosomes do humans have in their body cells?

3. What are genes?

4. Name three things that increase the chance of mutations.

5. In what circumstances are mutations inherited?

6. What is Down's syndrome?

Examiner's Top Tip
Make sure you learn how the chances of mutation can be increased.

6. A chromosome mutation, an extra chromosome number 21
5. If they occur in reproductive cells
4. Ionising radiation, X-rays and UV light; certain chemicals
3. Sections of DNA that code for a particular characteristic
2. 46
1. Chromosomes

SELECTIVE BREEDING

ARTIFICIAL SELECTION

People are always trying to breed animals and plants with special characteristics.
For example, a fast racehorse or a cow that produces lots of milk.
<u>The procedures involved in artificial selection are:</u>
• Select the individuals with the best characteristics.
• Breed them together using sexual reproduction.
• Hopefully some of the offspring will have inherited some of the desirable features; the best offspring are selected and are bred together.
• This is repeated over generations until the offspring have all the desired characteristics.
The following is an example of selective breeding to produce a large, tasty strawberry:
• The small, but nice tasting strawberry was allowed to breed with the large but tasteless strawberry to produce seeds.
• The seeds were allowed to grow.
• Some of the new strawberries were large and tasty.
• These were bred together to produce seeds that developed into all large and tasty strawberries.

small and tasty

large but tasteless

SELECTIVE BREEDING IN ANIMALS

The farm pig has been selectively bred over the years from a wild pig.
The features that have been bred in are:
• *less hair*
• *a quieter temperament*
• *fatter*
Can you think why?
• *Cows have been selectively bred to produce a greater quantity of milk.*
• *Beef cattle have been bred to produce better meat.*
The problems with only breeding from the best cows and bulls is that the cows can only give birth once a year. New techniques have therefore been developed.

wild pig

modern farm pig

SELECTIVE BREEDING IN PLANTS

· Selectively bred individuals may not always produce the desired characteristics as <u>sexual</u> <u>reproduction</u> <u>always</u> <u>produces</u> <u>variation</u>.
· With plants this can be overcome by producing <u>clones</u>.

· <u>Clones</u> <u>are</u> <u>genetically</u> <u>identical</u> <u>individuals</u>.
· To produce clones, asexual reproduction is needed.
· Many plants reproduce asexually on their own, for example strawberry plants, which produce runners.

PROBLEMS WITH SELECTIVE BREEDING

• The problem is a <u>**reduction**</u> **in** <u>**the**</u> <u>**number**</u> **of** <u>**alleles**</u> **in** **a** <u>**population**</u>.
• If animals or plants are continually bred from the same best animals or plants, the animals and plants will all be very similar.
• If there is a change in the environment, the new animals and plants may not be able to cope with it.
• There may be no alleles left to breed new varieties of plants and animals selectively.
• It is important to keep wild varieties alive, to maintain species variation.

EMBRYO TRANSPLANTS

The process is as follows:
- Sperm is taken from the best bull.
- The best cow is given hormones to stimulate the production of lots of eggs.
- The eggs are removed from the cow and are fertilised in a petri dish.
- The embryos are allowed to develop, but are then split apart to form clones before they become specialised.
- The embryos are then implanted into other cows called surrogates, where they grow into offspring.
- Advantages: the sperms and the eggs can be frozen to be used at a later date; a large number of offspring can be produced from one bull and one cow.

sperm taken from best bulls

fertilisation in petri dish

eggs taken from best cows

fertilised egg grows into embryo

embryos transplanted into surrogate cows

TISSUE CULTURE AND CUTTINGS

- Gardeners can produce new, identical plants by taking cuttings from an original parent plant.
- The plants are dipped in rooting powder containing hormones and are kept in a damp atmosphere to grow into new plants. The new plants would be clones.

trim off lower leaves and make a slanting cut just below a leaf stalk

roots grow from stem after several weeks in water and hormone powder

- Tissue culture is a technique used by commercial plant breeders.
- They take just a few plant cells and grow a new plant from them, using a special growth medium containing hormones.
- The advantages are that new plants can be grown quickly and cheaply all year round, with special properties such as resistance to diseases.
- Plants also reproduce sexually, attracting insects for pollination.
- The resulting plants show variation.
- This is very important because if they only produced clones and a new disease developed, it would kill the one clone and the others in that species, as they would be all the same.

ARTIFICIAL INSEMINATION

This is a technique that involves taking sperms from a male and artificially inseminating the best females with the sperm. The sperm can be frozen. In this way farmers don't have to keep the best male – it can even be in another country.

Examiner's Top Tip
Learn the four steps involved in selective breeding and remember it is also called artificial selection.

Examiner's Top Tip
Make sure you can list the advantages of selective breeding and also the disadvantages.

QUICK TEST

1. What is selective breeding?
2. What is the difference between artificial selection and natural selection?
3. What are clones?
4. Name two techniques used in the selective breeding of animals, particularly cows and bulls.
5. Name two methods of selective breeding in plants.
6. What is the main disadvantage of selective breeding?

1. Breeding animals or plants to produce the best offspring
2. In artificial selection, humans do the selecting, rather than nature
3. Genetically identical individuals
4. Embryo transplants and artificial insemination
5. Taking cuttings and tissue culture
6. A reduction in the number of alleles

EXAM QUESTIONS - Use the questions to test your progress. Check your answers on page 95.

1. How long does the menstrual cycle last?

....

2. Where is information stored about your inherited characteristics?

....

3. What causes variation in plants and animals?

....

4. Give an example of:
a) discontinuous variation..
b) continuous variation ....

5. Give an example of how the environment can affect the appearance of a plant.

....

6. How many chromosomes does a human sperm cell have?

....

7. How can the chance of mutations be increased?

....

8. What is selective breeding?

....

9. What is the difference between asexual and sexual reproduction?

....

10. Cystic fibrosis is an example of a genetic disease. What organ does it affect?

....

11. How is cystic fibrosis inherited?

....

12. Give an example of another inherited disease that is dominant.

....

13. Give an example of a recessive inherited disease other than cystic fibrosis.

....

14. What are gametes?

....

15. What chromosomes must a female have?

....

16. Describe how you would obtain a large, tasty strawberry from a small, tasty strawberry and a large, tasteless strawberry.

....

17. Where are genes found?

....

18. If an allele is dominant, what does this mean?

....

19. What is the difference between genotype and phenotype?

....

20. What does mitosis produce, and where does it occur?

....

21. Where does meiosis occur and what does it produce?

....

22. How many chromosomes does a human body cell have before it divides by mitosis?

....

23. How many chromosomes will a body cell have after it has divided by mitosis?

....

24. Which hormones are involved in the menstrual cycle?

....

25. Which part of the brain is involved in the control of the menstrual cycle?

....

26. Which hormone begins the menstrual cycle?

....

27. Complete the Punnett square to show the outcome of a cross between two heterozygous parents for eye colour

Parents	Mother		Father
phenotype	brown		brown
genotype	Bb	X	Bb
gametes	B or b		b or B
(sperm and eggs)	(eggs)		(sperm)

	B	b
b		
B		

28. What organisms do we use for genetic engineering?

....

29. How do we use genetic engineering to help diabetics?

....

30. What do genes code for and what is the process called?

....

How did you do?

1–7	correct	....start again
8–15	correct	....getting there
16–22	correct	....good work
23–30	correct	....excellent

- **Carbon dioxide is a rare atmospheric gas.**
- **It makes up 0.03% of the atmosphere.**
- **This amount should stay the same, as the carbon is recycled.**
- **The amount of carbon released into the atmosphere balances the amount absorbed by plants.**

PHOTOSYNTHESIS

Plants absorb carbon dioxide from the air. They use the carbon to make carbohydrates, proteins and fats using the **Sun** as an energy source.

FEEDING

Animals eat plants and so the carbon gets into their bodies.

DEATH AND DECAY

Plants and animals die and produce waste. The carbon is released into the soil.

DECOMPOSERS

Bacteria and fungi present in the soil break down dead matter, urine and faeces, which contain carbon.
Bacteria and fungi release carbon dioxide when they respire.

DEATH BUT NO DECAY

Sometimes plants and animals die, but do not decay. Heat and pressure gradually, over millions of years, produce fossil fuels.

Examiner's Top Tip
Remember, there is only one way carbon enters the cycle (photosynthesis) and two ways it is released back into the atmosphere (respiration and combustion).

THE CARBON CYCLE

DECOMPOSITION

- <u>Bacteria</u> and <u>fungi</u> are <u>decomposers</u>. They break down dead material.
- As well as helping to recycle carbon into the atmosphere, decomposers also recycle other <u>nutrients</u> into the soil.
- Plants take up these nutrients dissolved in water during photosynthesis.
- Animals eat plants; animals and plants eventually die, and the whole process begins again.
- Decomposition happens everywhere in nature, and also in compost heaps and sewage works.
- The ideal conditions for decomposition are <u>warmth</u>, <u>moisture</u> and <u>oxygen</u>.

Examiner's Top Tip
The carbon cycle in the exam may look slightly different, so make sure you learn the processes involved.

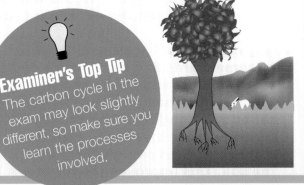

RESPIRATION

Plants, animals and decomposers respire. Respiration releases carbon dioxide back into the atmosphere.

BURNING AND COMBUSTION

The burning of fossil fuels (coal, oil and gas) releases carbon dioxide into the atmosphere.

FOSSIL FUELS

Coal is formed from plants; oil and gas are formed from animals.

QUICK TEST

1. Name the process that absorbs carbon dioxide from the air.
2. What are the two ways that carbon is released back into the air?
3. How do animals get the carbon into their bodies?
4. Name the three types of organisms that carry out respiration.
5. What is decomposition?
6. What organisms are involved in decomposition?
7. What happens to the bodies of animals and plants that do not decay?
8. What are the ideal conditions for decomposition of dead matter to occur?
9. What do the plants do with the carbon they absorb?
10. How much carbon dioxide is present in the atmosphere?

1. Photosynthesis
2. Respiration and burning/combustion
3. Eat plants
4. Animals, plants and decomposers
5. Breaking down of dead material
6. Decomposers, bacteria and fungi
7. Turned into fossil fuels
8. Warmth moisture and oxygen
9. Turn it into carbohydrates, proteins and fats
10. 0.03%

NITROGEN GAS IS CHANGED INTO NITRATES IN THE NITROGEN CYCLE

Follow the numbers on the diagram to see what happens at each stage.

1. Nitrogen is in the air.

2. **Lightning** causes nitrogen and oxygen to combine to form nitrogen oxides. These dissolve in rain and are washed into the soil to form **nitrates**.

lightning

3. **Nitrogen-fixing bacteria** in the soil convert nitrogen from the air into **nitrates**.

4. **Nitrogen-fixing bacteria** in the roots of some plants like peas, beans and clover also change nitrogen into **nitrates**. The bacteria form lumps called **root nodules**. Sometimes farmers grow these leguminous plants and then plough them into the soil. This improves the nitrate content of the soil so that other crops can be grown successfully.

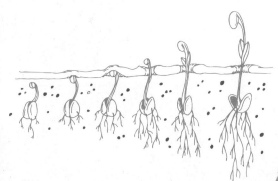

nitrogen-fixing bacteria in root nodules

5. Fertilisers generated from the Haber process can be added to the soil to improve the **nitrate** content.

6. Plants take up the **nitrates** from the soil and convert them into **proteins**.

7. **Animals eat the plants**, taking the proteins into their bodies where it becomes part of the animals' proteins.

8. Animals and plants produce waste.

9. Animals and plants eventually die and their bodies decay.

10. **Detritivores**, such as worms, maggots and woodlice, feed on this dead and decaying material and make it easier for decomposers to break down.

remains

11. **Decomposers**, such as fungi and bacteria, turn this material into ammonium compounds that contain nitrogen.

12. **Nitrifying bacteria** in the soil change **ammonia into nitrates**.

13. **Nitrates** can be washed out of the soil before plants take them up. This is called **leaching** and can have serious consequences for rivers and streams.

Examiner's Top Tip
Make sure you know the 3 types of bacteria involved in the nitrogen cycle.

denitrifying bacteria in waterlogged soil

14. **Denitrifying bacteria** live in waterlogged soils; they can change **nitrates** back into ammonia and nitrogen gas that is returned to the atmosphere.

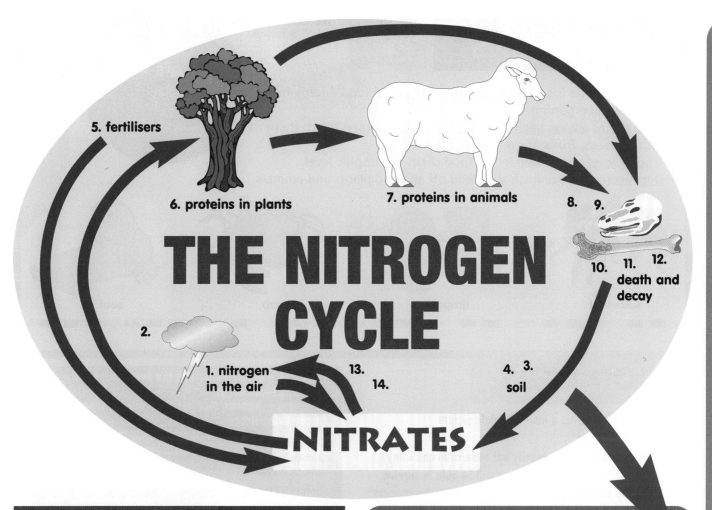

5. fertilisers

6. proteins in plants

7. proteins in animals

8. 9.

10. 11. 12. death and decay

THE NITROGEN CYCLE

2.

1. nitrogen in the air

13.

14.

4. 3. soil

NITRATES

SUMMARY

- Lightning, artificial fertilisers and nitrogen-fixing bacteria in the soil and in root nodules convert nitrogen into nitrates.
- Nitrifying bacteria in the soil change animal waste and dead remains into ammonia and eventually nitrates.
- Denitrifying bacteria live in water-logged soils and convert nitrates back into nitrogen.

- The atmosphere contains 78% nitrogen gas.
- Nitrogen is an important element needed for making <u>proteins</u>.
- Plants and animals cannot use nitrogen in this form.
- It has to be <u>converted</u> <u>to</u> <u>nitrates</u> before plants can use it to make proteins.
- We get the proteins into our bodies by eating plants.
- It's a continuous cycle, so can begin anywhere.

Examiner's Top Tip
There are four ways in which nitrogen is converted into nitrates in the soil and two ways nitrates are taken out of the soil.

QUICK TEST

1. What do plants need nitrogen for?
2. What does nitrogen have to be converted into before it is used?
3. What part does lightning play in the nitrogen cycle?
4. Where are nitrogen-fixing bacteria found?
5. Where are nitrifying bacteria found and what do they do?
6. What are denitrifying bacteria?
7. What is another way that nitrates are taken out of the soil?
8. Why are leguminous plants good for the soil?
9. What happens to the waste and dead remains of plants and animals?
10. Where do animals get their proteins from?

10. By eating plants
9. Broken down by detritivores and decomposers into ammonia compounds
8. They contain nitrifying bacteria in their roots nodules; these convert nitrogen into nitrates
7. Taken up by plants, dissolved in water
6. Bacteria that are present in waterlogged soils that convert nitrates back into nitrogen
5. In the soil, they convert ammonia from waste and remains into nitrates
4. In the soil and root nodules of some plants
3. Causes nitrogen and oxygen to combine at high temperature to form nitrogen oxides which dissolve in rain and form nitrates in the soil
2. Nitrates
1. For making proteins

77

FOOD CHAINS

- The <u>arrows</u> in a food chain show the transfer of food energy from organism to organism.
- <u>Food chains always begin with the Sun, then a green plant</u>; these can include seeds, fruits or even dead leaves.
- <u>Trophic level</u> – each step in the food chain is a trophic level.
- <u>Detritivores</u> – are animals that feed off all dead plants and animals.

Examiner's Top Tip
Learn all the definitions involved in food chains and webs.

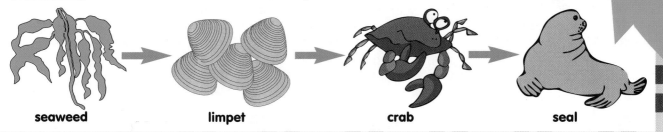

seaweed   limpet   crab   seal

FOOD WEBS

- A food web gives us a more complete picture of who eats what.
- Most animals in a community eat more than one thing. If one kind of food runs out, they will be able to survive by eating something else.
- Food webs are made up of many food chains linked together.
- Food chains can be drawn for any environment.

<u>Producers</u> – green plants use the Sun's energy to produce food energy
<u>Consumers</u> – animals that get their energy from eating other living things
<u>Primary consumers</u> – animals that eat the producers
<u>Secondary consumers</u> – animals that eat the primary consumers
<u>Tertiary consumers</u> – animals that eat the secondary consumers
<u>Herbivores</u> – animals that only eat plants
<u>Carnivores</u> – animals that only eat animals
<u>Top carnivores</u> – animals that are not eaten by anything else except decomposers after they die

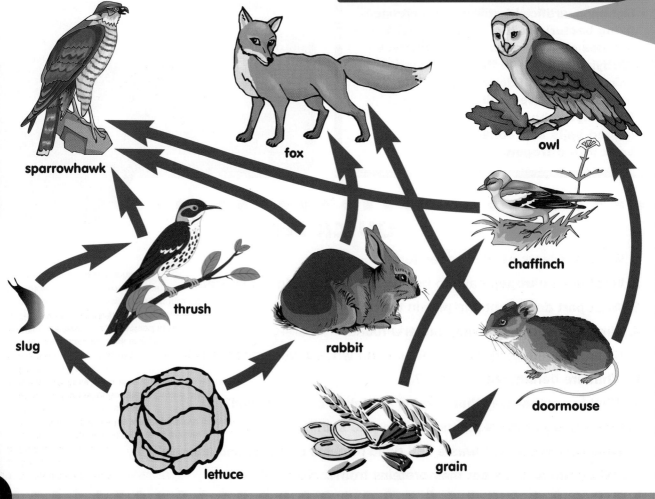

sparrowhawk   fox   owl   chaffinch   thrush   rabbit   doormouse   slug   lettuce   grain

TYPICAL EXAM QUESTIONS

- Exam questions often focus on what would happen to a food web if an animal was removed by disease or other factors.
- In the food web (below left), what would happen if the rabbit was removed?
1. The amount of lettuce would increase at first as only the slugs will be eating it.
2. The slugs would increase as there are more lettuces; the thrushes would increase at first because there are more slugs, but then the thrushes would get eaten by the sparrowhawks.
3. The number of foxes and sparrowhawks would struggle at first with no rabbits, but then they would eat more doormice and thrushes.
4. The doormice would decrease and so would the number of owls.
- It all looks incredibly complicated, but you just have to think it through for every animal.
- Look at who would get eaten, who would go hungry, what would they do about it and what effect it would have on the other animals in the food web.

FOOD CHAINS AND WEBS

- Food chains and webs <u>begin</u> <u>with</u> <u>energy</u> <u>from</u> <u>the</u> <u>Sun</u>.
- A <u>food</u> <u>chain</u> shows us what eats what in a community.
- A <u>food</u> <u>web</u> is made up of interconnected food chains.

grass   rabbit   fox

QUICK TEST

1. What is a producer?

2. What is a consumer?

3. Name the producers in the food web opposite.

4. Name the primary consumers in the food web opposite.

5. How many top carnivores are there in the food web opposite?

6. What are herbivores?

7. What are carnivores?

8. How many carnivores are there in the food web opposite?

9. What part does the Sun play in food chains and webs?

10. What do the arrows in food webs and food chains show?

1. A plant that produces food from the Sun's energy
2. An animal that eats other plants and animals
3. Lettuce and grain
4. Slug, rabbit, chaffinch and dormouse
5. Three
6. Animals that eat only plants
7. Animals that eat only other animals
8. Four
9. It provides the energy to start a food chain
10. The transfer of food energy between organisms

PYRAMIDS OF NUMBERS AND BIOMASS

- A pyramid of numbers tells us the number of organisms involved in a food chain.
- A pyramid of biomass tells us the mass of the organisms involved in a food chain.

PYRAMIDS OF NUMBERS

- If we look at the information a food chain tells us, it is simply who eats who.
- A pyramid of numbers tells us how many organisms are involved at each stage in the food chain.
- At each trophic level the number of organisms gets less.

blackbird
ladybirds
aphids
rosebush

- A rose bush and a grass plant count as one organism, but a rose bush can support more herbivores than a grass plant.

fox
rabbit
grass

- Sometimes a pyramid of numbers does not look like a pyramid at all as it doesn't take into account the size of the organisms.

fleas
fox
rabbits
lettuce

- In this pyramid of numbers the top carnivores are fleas that feed on single fox.

PYRAMIDS OF BIOMASS

- A biomass pyramid takes into account the size of an organism at each level.
- It looks at the mass of each organism.
- If we take the information from the pyramid of numbers and multiply it by the organism's mass, we get a pyramid shape again.

blackbird
ladybirds
aphids
rosebush

- A single rose bush weighs more than the aphids, and lots of aphids weigh more than the few ladybirds that feed on them.
- A blackbird weighs less than the many ladybirds it feeds on.

fleas
fox
rabbits
lettuce

- Even though there are a lot of fleas, they weigh less than the fox they feed on.
- The fox weighs less than the number of rabbits it eats, and the number of lettuces the rabbits eat weigh more than the rabbits.

WHERE DOES THE ENERGY GO?

- The 90% energy loss at each stage goes on life processes such as respiration.
- Respiration releases heat energy to the surroundings.
- Animals that are warm blooded use up a lot of energy in keeping warm, so they need to eat a lot more food.
- As you can see from the diagram of energy transfers in the cow, a lot of energy is lost in urine and faeces.
- Not all of the organism's body mass is eaten.

LOSS OF ENERGY IN FOOD CHAINS

- Food chains rarely have more than four or five links in them.
- This is because energy is lost along the way.
- The final organism is only getting a fraction of the energy that was produced at the beginning of the food chain.
- Plants absorb energy from the Sun.
- Only a small fraction of this energy is converted into glucose during photosynthesis.
- Some energy is lost to decomposers as plants shed their leaves, seeds or fruit.
- The plant uses some energy during respiration and growth.
- The plant's biomass increases, which provides food for the herbivores.
- Only approximately 10% of the original energy from the Sun is passed on to the primary consumer in the plant's biomass.
- The primary consumer also has energy losses and only approximately 10% of its total energy intake is passed on to the secondary consumer.

Sun → primary consumers

released in respiration or lost to decomposers

ENERGY FLOW THROUGH A PRODUCER

30% lost as heat in respiration

60% lost in urine and faeces

energy consumed

10% in growth

ENERGY FLOW THROUGH A CONSUMER

EFFICIENCY OF FOOD PRODUCTION

As so much energy is lost along food chains, we must look at ways to improve the efficiency of food production and reduce losses.

There are two ways this can be done:

- reduce the number of stages in the food chain – it is more energy efficient to eat plant produce than meat.
- rear animals intensively – restrict their movement, keep them warm and they won't need feeding as much.

However, the last way is not a decent way to keep animals.

Examiner's Top Tip
A common exam question is about the energy losses in food chains and how to reduce them, so make sure you learn them thoroughly.

QUICK TEST

1. Why do food chains only have four or five links?

2. What do pyramids of numbers show?

3. What don't they take into account?

4. What pyramids can be drawn using the mass of animals and plants?

5. Where does a plant get its original energy source?

6. Approximately how much energy is passed on from the producer to the consumer?

7. Why do warm-blooded animals need to eat a lot of food?

8. In animals, where does most of the energy go?

9. List the ways that energy is lost in food chains.

10. How can we reduce energy loss in food chains?

Examiner's Top Tip
Questions in the exam may provide you with actual numbers or masses of the organisms involved in food chains, make sure you draw them to scale.

1. Energy is lost along the way
2. The numbers of organisms involved in a food chain
3. The size of the organisms
4. Pyramids of biomass
5. The Sun
6. 10%
7. They need to keep warm
8. Lost in urine and faeces
9. Respiration, heat, waste and parts of the body not eaten
10. Eat lower down in the food chain and intensively rear animals

81

THE THEORY OF EVOLUTION

Religious theories are based on the need for a 'creator' for all life to exist on Earth, but there are other theories.

Charles Darwin, a British naturalist, first put forward his theory about 140 years ago. Darwin visited the Galapagos Islands off the coast of South America and made a number of observations that led to his theory of evolution:

- *Organisms produce more offspring than can possibly survive.*
- *Population numbers remain fairly constant despite this.*
- *All organisms in a species show variation.*
- *Some of these variations are inherited.*

He also concluded from these observations that since there were more offspring produced than could survive, there must be a struggle for existence.

This led to the strongest and fittest offspring surviving and passing on their genes to their offspring.

organisms produce a large number of offspring

in any species there is variation between individuals

there is a struggle for existence

organisms with useful characteristics are more likely to survive and pass them on to the next generation

- *This is sometimes called the <u>survival of the fittest</u> or <u>natural selection</u>.*

A man called Lamarck suggested another theory. His theory was that animals evolved features according to how much they used them.

His idea was that giraffes, for example, stretched their necks reaching for food, their offspring had longer necks which they stretched further. This led to the modern long-necked giraffe.

- *<u>Eventually Darwin's theory was accepted, although religious beliefs should also be respected</u>.*

NATURAL SELECTION

- Darwin stated that the process of natural selection was the basis for evolution.
- A species is defined as a group of living things that are able to breed together and produce fertile offspring.
- Within a species there is variation between individuals.
- Changes in the environment may affect some individuals and not others.
- **<u>Only those who can adapt to suit their new environment survive to breed and pass on their advantageous genes</u>.**
- Other factors that tend to prevent all offspring surviving are: competition for food; predators and disease.
- Eventually it seems that nature has decided which individuals should survive and breed. There is a **<u>'survival of the fittest'</u>**.

NATURAL SELECTION IN ACTION

- An example of the environment causing changes in a species is the <u>peppered moth</u>.
- They live in woodlands on lichen-covered trees.
- There are two types of peppered moth: a light, speckled form and a dark form.
- The dark-coloured moth was caused by a mutation and was usually eaten by predators.
- In the 1850s the dark type of moth was rare, but pollution from factories started to blacken tree trunks.
- The dark moth was then at an advantage because it was camouflaged.
- In 1895 most of the population of moths were dark.
- In cleaner areas the light moth had an advantage against predators, so it still survived to breed.

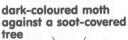

dark-coloured moth against a soot-covered tree

the pale moth is at a disadvantage in polluted areas

MUTATIONS

- Mutations that occur in genes are usually harmful, but in the case of the peppered moth the mutation turned out to be useful when the environment changed.
- Over a long period of time, gradual changes (mutations) may result in totally new species being formed.
- This brings us back to the theory of evolution, that all species evolved from a common ancestor that existed billions of years ago.
- Those species that were unable to adapt to their surroundings became <u>extinct</u>.

EVOLUTION

- <u>Evolution</u> is all about change and improvement from simple life forms.
- The <u>theory</u> of <u>evolution</u> states that all living things that exist today or existed, evolved from simple life forms three billion years ago.
- <u>Natural selection</u> is the process that causes evolution.
- <u>Fossils</u> provide the evidence for evolution.

FOSSILS

- <u>Fossils are the remains of dead organisms that lived millions of years ago. They are found in rocks.</u>
- Most dead organisms decay and disintegrate, but the following are ways that fossils can be formed:
1. The hard parts of animals that don't decay form into a rock.
2. Minerals which preserve their shape gradually replace the softer parts of animals that decay very slowly.
3. Fossils are formed in areas where one or more of the conditions needed for decay are absent; for example, areas where there is <u>no oxygen</u>, <u>moisture or warmth</u>.
- Fossils provide evidence for evolution.
- They are preserved in rock, with generally the younger fossils being found nearer the surface.

- The evolution of the horse is clearly shown by fossils.
- Look at the changes in their size and feet caused by the changes in the environment over the years.
- Natural selection has operated to produce the modern horse.

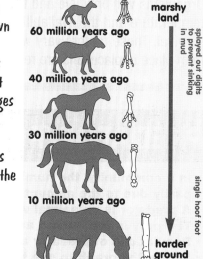

marshy land

splayed out digits to prevent sinking in mud

60 million years ago

40 million years ago

30 million years ago

10 million years ago

single hoof foot

1 million years ago   modern horse

harder ground

QUICK TEST

1. Who were the two scientists that put forward their theories of evolution?

2. Whose theory was gradually accepted?

3. What prevents all the organisms in a species surviving?

4. Which individuals would survive a change in the environment?

5. What process causes evolution?

6. Why did the feet of the modern horse change?

7. Which type of peppered moth would survive in industrial regions and why?

8. What conditions need to be absent for fossils to be formed?

8. Oxygen, moisture and warmth
7. The dark form; camouflaged against predators
6. The ground changed from marshy land to hard ground
5. Natural selection/survival of the fittest
4. The best adapted
3. Environmental change, competition, disease and predators
2. Darwin's
1. Darwin and Lamarck

ADAPTATION AND COMPETITION

- A <u>habitat</u> is where an organism lives; it has the conditions needed for it to survive.
- A <u>community</u> consists of living things in the <u>habitat</u>.
- Each <u>community</u> is made up of different <u>populations</u> of <u>animals</u> <u>and</u> <u>plants</u>.
- Each <u>population</u> is adapted to live in that particular habitat.

SIZES OF POPULATIONS

Population numbers cannot keep growing out of control; factors that keep the population from becoming too large are called <u>limiting</u> <u>factors</u>. The factors that affect the size of a population are:
- Amount of food and water available
- Predators or grazing – who may eat the animal or plant
- Disease
- Climate, temperature, floods, droughts and storms
- Competition for space, mates, light, food and water
- Human activity such as pollution or destruction of habitats

Organisms will only live and reproduce where conditions are suitable; the amount of light, the temperature and the availability of food and water will affect the organisms and are essential for their survival. These factors vary with time of day and time of year; this helps explain why organisms vary from place to place, and are restricted to certain habitats. Organisms have adapted to live in certain areas.

PREDATOR/PREY GRAPHS

- **In a community, the number of animals stays fairly constant; this is partly due to the amount of food limiting the size of the populations.**
- **A <u>predator</u> is an animal which hunts and kills another animal.**
- **The <u>prey</u> is the hunted animal.**
- **Populations of predator and prey go in cycles.**
- **Follow the graph to see how the numbers of prey affect the numbers of predators and vice versa.**

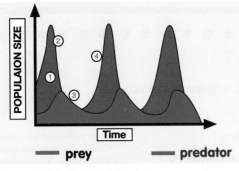

prey — predator

1. If the population of prey increases, there is more food for the predator, so its numbers increase.
2. This causes the number of prey to decrease as they are eaten.
3. This causes the number of predators to decrease, as there is not enough food.
4. If the predator numbers fall, the prey numbers can increase again, as they are not being eaten, and so on.

- <u>Predators</u> <u>have</u> <u>adapted</u> to survive by being strong, agile and fast. They have good vision and a camouflaged body. They also tend to hunt in packs, have a variety of prey, and often hunt the young, sick and old.
- <u>Prey</u> <u>have</u> <u>also</u> <u>adapted</u>; the best adapted escape and breed.
- Adaptations of prey include: being able to run, swim and fly fast; they stay in large groups; they have a horrible taste; warning colours and camouflage.

COMPETITION

- As populations grow, there may be overcrowding and limited resources to support the growing numbers.
- Animals have to compete for <u>space</u>, <u>food</u> and <u>water</u> in their struggle to survive.
- Only the strongest will live, leading to the survival of the fittest.
- Plants compete for <u>space</u>, <u>light</u>, <u>water</u> and <u>nutrients</u>.
- The weed is a very successful competitor; see the diagram to see how.

grows quickly and flowers twice a year

resistant to many weedkillers

grows quickly on bare soil

roots produce chemicals that stop other plants growing

produces many seeds which are spread by the wind

seeds germinate rapidly

leaves spread out over ground

deep roots which are difficult to remove

ADAPTATION

You never see a polar bear in the desert or a camel at the North Pole. This is because they have not <u>adapted</u> to live there. They have adapted to live where they do; they have <u>special</u> <u>features</u> that help them survive.

A polar bear lives in cold, arctic regions of the world; it has many features that enable it to survive:

- It has a <u>thick</u> <u>coat</u> to keep in body heat, as well as a <u>layer</u> <u>of</u> <u>blubber</u> for insulation.
- Its coat is <u>white</u> so that it can blend into its surroundings.
- Its <u>fur</u> <u>is</u> <u>greasy</u> so that it doesn't hold water after swimming. This prevents cooling by evaporation.
- A polar bear has <u>big</u> <u>feet</u> to spread its weight on snow and ice; it also has big, <u>sharp</u> <u>claws</u> to catch fish.
- It is a <u>good</u> <u>swimmer</u> and <u>runner</u> to catch prey.
- The shape of a polar bear is <u>compact</u> even though it is large. This keeps the surface area to a minimum to reduce loss of body heat.

A camel has features that enable it to survive in the hot deserts of the world:

- The camel has an ability to <u>drink</u> a lot of water and <u>store</u> it.
- It loses very little water as it produces <u>little</u> <u>urine</u> and it can cope with big changes in temperature, so there <u>is</u> <u>no</u> <u>need</u> <u>for</u> <u>sweating</u>.
- All fat is stored in the humps, so there is <u>no insulation layer</u>.
- Its <u>sandy</u> colour provides <u>camouflage</u>.
- It has a <u>large</u> <u>surface</u> <u>area</u> to enable it to lose heat.

In a community, the animal or plant best adapted to its surroundings will survive.

QUICK TEST

1. What is a habitat?
2. Define the word community.
3. What makes up a population?
4. What things do animals compete for?
5. What things do plants compete for?
6. If the number of prey increases what will happen to the number of predators?
7. Why do the numbers of prey and predators in a community stay fairly constant?
8. How does the polar bear's coat help it survive in the arctic?
9. How does the camel's large surface area help it survive in the desert?
10. What factor determines whether animals or plants survive in their environments?

10. Only the best adapted organism's will survive
9. Enables it to lose heat to its surroundings
8. White (camouflage), thick (insulation) and greasy (does not hold water)
7. Because of the predator – prey cycle
6. Also increase at first
5. Light, space, water and nutrients
4. Food, water and space
3. Animals or plants
2. The living things in a habitat
1. Where an organism lives

EFFECTS ON THE ENVIRONMENT

- Humans are using up the Earth's resources, including fossil fuels, at an alarming rate.
- Burning fossil fuels contributes to: <u>acid rain</u> and the <u>Greenhouse Effect</u> by releasing harmful gases into the air.
- Gases such as sulphur dioxide dissolve in rain and make it acidic.
- Acid rain damages wildlife and pollutes rivers and lakes.

- Carbon dioxide is also released into the atmosphere by burning fossil fuels.
- This gas traps heat inside the Earth's atmosphere and causes the temperature of the Earth to increase. This is known as the Greenhouse Effect.

FERTILISERS

- Plants need nutrients to grow, which they take up from the soil.
- With intensive farming methods, nutrients are quickly used up, so the farmer has to replace them with artificial fertilisers.
- Fertilisers enable farmers to produce more crops in a smaller area of land, and can reduce the need to destroy the countryside for extra space.
- However, problems – particularly <u>eutrophication</u> – are caused by the use of fertilisers.

HUMAN INFLUENCE ON THE ENVIRONMENT

- Improvements in agriculture, health and medicine have meant a dramatic <u>rise in human populations</u>.
- An increase in population size leads to an increase in pollution and higher demands on the world's resources.

DEFORESTATION

- In the UK there are already not many forests left.
- In under-developed countries people are chopping down forests to provide timber or space for agriculture, to try to provide for the growing numbers of people.
- This causes several problems to the environment.
- Burning this timber <u>increases</u> the <u>level of carbon dioxide in the air.</u>
- Forests absorb carbon dioxide in the air and provide us with oxygen.
- Chopping down trees leads to <u>soil erosion</u> as the soil is exposed to rain and wind.
- The trees evaporate water into the air and without them there will be a <u>decrease in rainfall</u>.
- Destroying forests also <u>destroys many different animal and plant habitats</u>.

EUTROPHICATION

- If too much fertiliser is added to the soil and it rains soon afterwards, the fertiliser finds its way into rivers and lakes.
- This causes the water plants to grow more quickly and cover the surface of the water.
- More water plants means more competition for light and some plants die.
- Microbes (bacteria and fungi) break down the dead plants and use up oxygen for respiration.
- This reduces the amount of oxygen available for animals and they die of suffocation.
- <u>Untreated sewage</u> pumped into rivers and streams also causes eutrophication.

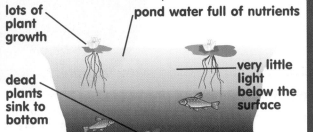

lots of plant growth

pond water full of nutrients

dead plants sink to bottom

very little light below the surface

microbes break down dead plants and use up oxygen through respiration

fish and other animals suffocate

INTENSIVE FARMING

Farming has had to become more intensive to try and provide more food from a given area of land.
Intensive farming can produce more food, but it has its problems.
Many people regard intensive farming of animals as cruel.
In order to produce more food from the land, <u>fertilisers</u> <u>and</u> <u>pesticides</u> are needed.

DESTRUCTION OF THE LAND

An increase in industry has led to the need to take over the land, which destroys wildlife and causes pollution.

we use land for building

dumping our rubbish

getting raw materials

farming to feed the world

Examiner's Top Tip
Learn arguments for and against intensive farming.

PESTICIDES

- *Pesticides are used to kill the insects that damage crops.*
- *They also kill harmless insects, which can cause a shortage of food for insect-eating birds.*
- *There is always the danger that pesticides can get washed into rivers and lakes and end up in our food chains.*
- *This was the case in the 1960s when a pesticide called DDT got into the food chain and threatened populations of animals.*

p.p.m. = parts per million

the pesticide DDT is magnified by the time it enters the grebe's body

(1600 p.p.m. of DDT) grebes

(250 p.p.m. of DDT) fish

(5 p.p.m. of DDT) plankton

(0.02 p.p.m. of DDT) water

WHAT CAN BE DONE?

- The problems will get worse unless people can learn to limit their needs and therefore prevent the destruction of our planet.
- Intensive farming does produce quality food, more than enough to supply people's needs in Europe, but it also creates many problems.
- A possible solution to some of the problems is organic farming.
- Organic farming produces less food per area of land and can be expensive, but it attempts to leave the countryside as it is and is kinder to animals.
- Organic farming uses manure as a fertiliser, sets aside land to allow wild plants and animals to flourish and uses biological control of pests.
- Biological control of pests is the use of other animals to eat pests; it is not as effective but produces no harmful effects.
- Another method that could be used to reduce harm to the environment is the use of greenhouses to grow food efficiently and out of season.
- We can also look at developing alternative energy sources, such as solar power and wind energy.
- This will help conserve the world's rapidly diminishing fossil fuel supply.

Examiner's Top Tip
Don't forget about acid rain and the Greenhouse Effect as problems have intensified with the growing human population.

QUICK TEST

1. Why has the human population increased in last few hundred years?
2. What problems can the use of pesticides cause?
3. How do fertilisers get into rivers and lakes?
4. What is the name of the process whereby fertilisers kill animals and plants in water?
5. Name four ways that humans reduce the amount of land.
6. What is deforestation?
7. How does deforestation contribute to the Greenhouse Effect?
8. What other problems does deforestation cause?
9. What could be used as an alternative to fertilisers?
10. Name ways in which we can reduce harmful effects on the environment.

1. Improved health care, medicine and agriculture
2. Problems in food chains
3. Washed in by rainwater
4. Eutrophication
5. Building, farming, dumping rubbish and quarrying
6. Cutting down of trees and forests
7. Less carbon dioxide is being absorbed
8. Soil erosion, less rainfall and the destruction of habitats
9. Manure
10. Organic farming, greenhouses and alternative energy sources

ACID RAIN, POLLUTION AND THE GREENHOUSE EFFECT

POLLUTION

- A pollutant is a substance that harms living things – animals, plants or humans.
- Pollutants can spread through the air, water and soil.

ACID RAIN

- Burning fossil fuels is the main cause of acid rain, so cars and power stations are the main culprits.
- The gases sulphur dioxide and various nitrogen oxides are released.
- They dissolve in water vapour in the clouds and fall as acid rain.

THE GREENHOUSE EFFECT

- Evidence is being collected that shows that the world is warming up.
- This is called global warming and is caused by the Greenhouse Effect.
- The temperature of the Earth is kept in balance by the heat we get from the Sun and the heat that is radiated back into the atmosphere.
- Carbon dioxide and water vapour act like an insulating layer (like glass in a greenhouse) and trap and keep some of the heat from the Sun.
- This is natural global warming and it provides enough heat for living things.
- The levels of carbon dioxide are increasing because of the burning of fossil fuels and the cutting down of trees (which absorb carbon dioxide).
- This increase in carbon dioxide is trapping too much heat and the Earth's temperature is slowly rising above normal.
- Methane gas is also contributing to the Greenhouse Effect.
- Methane is produced naturally from cattle waste and rice fields.

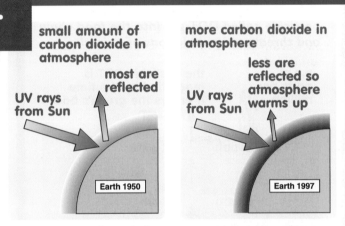

small amount of carbon dioxide in atmosphere

most are reflected

UV rays from Sun

Earth 1950

more carbon dioxide in atmosphere

less are reflected so atmosphere warms up

UV rays from Sun

Earth 1997

PROBLEMS CAUSED BY THE GREENHOUSE EFFECT

Changes in temperature could cause melting of the polar ice caps which in turn would cause raised sea levels. Serious flooding could result.

Plants may be killed by the warming and weeds may thrive as they grow well on extra carbon dioxide.

WHAT CAN BE DONE?

- We can look for alternative energy sources.
- Use unleaded petrol and reduce the need for cars.
- Use catalytic converters in cars to reduce emissions of harmful gases.
- Stop large scale deforestation because trees absorb carbon dioxide.

POLLUTION OF THE SOIL AND WATER

- Fertilisers and pesticides used on land can be washed into rivers and seas, causing damage to wildlife. Farm waste can have the same effects on wildlife (see topic on Human Influence on the Environment).
- Factory waste and sewage are often dumped at sea. Oil spillages at sea are a problem to marine life and seabirds and they also ruin beaches.

PROBLEMS CAUSED BY ACID RAIN

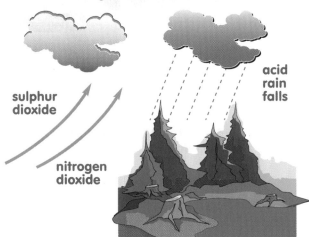

gases dissolve in water in clouds

sulphur dioxide

nitrogen dioxide

acid rain falls

- *Acid rain kills fish and trees and damages buildings, particularly ones made of limestone.*
- *Acid rain falls into lakes and poisons fish and birds.*

the effect acid rain can have on trees

POLLUTION OF THE AIR

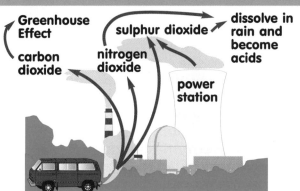

Greenhouse Effect

carbon dioxide

nitrogen dioxide

sulphur dioxide

dissolve in rain and become acids

power station

- Burning fossil fuels is the main cause of atmospheric pollution.
- Factories and cars use fossil fuels.
- They release carbon dioxide which contributes to the Greenhouse Effect and sulphur dioxide and nitrogen oxides which cause acid rain.
- Smoke from burning contains particles of soot (carbon) which can blacken buildings and collect on plants, preventing them from photosynthesising.
- The smoke from car engines also contains lead, which is known to cause damage to the nervous system if inhaled.
- CFCs (chlorofluorocarbons) are used in aerosols, refrigerators and in plastic foam.
- CFCs are also adding to the Greenhouse Effect as they cause the ozone layer to develop holes.
- This means that harmful UV rays are reaching the Earth's surface.
- This can lead to increased risk of skin cancer.
- Other gases now replace CFCs but the harmful effects of the CFCs already released will continue for some time.

QUICK TEST

1. What is the main cause of atmospheric pollution?
2. What gases cause acid rain?
3. What damage does acid rain cause?
4. Why is the Earth warming up?
5. Which two gases cause the Greenhouse Effect?
6. What does CFC stand for?
7. How can we reduce acid rain? Give two examples.
8. What causes holes in the ozone layer?
9. What harmful chemical damages the nervous system?
10. How can we reduce the effects that using cars has on the environment?

Examiner's Top Tip
The effects of acid rain and the greenhouse effect are common exam questions.

10. Use catalytic converters and unleaded petrol
9. Lead
8. CFCs
7. Use alternative energy sources and reduce the use of cars.
6. Chlorofluorocarbon
5. Carbon dioxide and methane
4. Increasing amounts of carbon dioxide and methane trapping too much heat
3. Damage to buildings, plants, fish and birds
2. Sulphur dioxide and nitrogen oxides
1. Burning fossil fuels

INDUSTRIAL FERMENTATION

- Microbes can be used to make chemicals on a large scale by growing them in a **fermenter**.
- To achieve the desired product as cheaply and efficiently as possible the following procedures need to be followed:
- **Temperature** needs to be maintained as microbes have an optimum temperature at which they and their enzymes work best.
- The water-cooled jacket removes excess heat from the process.
- **pH, oxygen and nutrient levels** need to be maintained; sensors in the fermentation vessel achieve this.
- The vessel must be kept **sterile**; the entry of unwanted substances would spoil the end product.
- The microbe needs to be kept supplied with **sufficient substrate (food)**.
- The vessel has a **stirrer** to ensure that the substrate and microbe are kept in close contact.

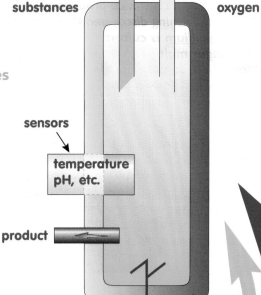

FOOD PRODUCTION

- Many food substances are produced by <u>fermentation</u>.
- They include yoghurt, bread, Marmite, beer, wine and mycoprotein.

MYCOPROTEIN
- Mycoprotein is a meat substitute marketed as Quorn.
- It is made from a fungus called <u>fusarium</u>.
- The fungus has the same texture as meat fibres but is tasteless and has to have flavour added.
- Mycoprotein is produced in a fermenter; the substrate for the fungus is glucose.
- In the fermentation vessel it grows extremely quickly; the advantages of mycoprotein are that it is high in protein and low in fat, with no cholesterol.
- It is a useful source of protein for vegetarians.

YOGHURT
- Yoghurt is produced by the fermentation of low-fat milk.
- Making yoghurt uses two species of bacteria, <u>Streptococcus</u> and <u>Lactobacillus</u>.
- The bacteria use the milk as a food source and decrease pH by producing lactic acid from the lactose sugar in milk.
- A pH of 4.4 causes the proteins to coagulate (thicken), so making thick yoghurt.

BEER
- Fermenting barley makes beer, it is flavoured with a substance called hops.
- Barley seeds germinate to make sugar.
- Yeast is added to the fermenter. It uses the sugar from barley to make alcohol and carbon dioxide. The alcohol is filtered, pasteurised to kill unwanted organisms and bottled.

Examiner's Top Tip
You don't need to learn the diagram of the fermenter, just the principles behind it.

Examiner's Top Tip
Learn examples of the uses of biotechnology.

OTHER PRODUCTS

PENICILLIN

- *Another useful substance produced by biotechnology is penicillin, an antibiotic.*
- *Alexander Fleming discovered that a mould called <u>Penicillium</u> destroyed bacteria.*
- *The <u>Penicillium</u> is cultured in a fermenter with its food source available.*
- *The fungus multiplies rapidly at first, but then runs out of energy.*
- *After about seven days the concentration of penicillin is at its maximum so it is removed and filtered to produce the antibiotic.*

BIOGAS

- *Biogas is mainly <u>methane</u>; microbes can digest waste products and produce methane gas during anaerobic respiration (respiration without oxygen).*
- *Waste products could be animal dung or waste from sugar factories and breweries; they all provide a food source for the microbes.*
- *Methane is used as a source of fuel for cooking or for generating electricity.*
- *A biogas generator is used to generate methane.*
- *Waste is fed in along with a food source; the microbes ferment the sugar into methane gas.*

methane gas

sugar waste and microbes

BIOTECHNOLOGY

making new fuels

brewing beer

making animal food

biotechnology has many uses today

baking bread

making antibiotics

PILLS

sewage disposal

- **Biotechnology has many uses in the modern world; its techniques have been used for a long time.**
- **Biotechnology means using plants, animals, bacteria and fungi (microbes) to produce useful substances or dispose of waste.**
- **Biotechnology may in the future solve the world's health, energy and food problems.**

QUICK TEST

1. Name two microbes.

2. What do the microbes need in order to reproduce?

3. Why is stirring necessary in the fermenter?

4. Name four food substances produced by biotechnology.

5. Name a substance that lowers pH and is produced when yoghurt is made.

6. Why is lowering the pH important?

7. Is penicillin produced from bacteria or a fungus?

8. Give an example of a biogas.

9. What is mycoprotein?

10. How can biotechnology help in the future?

1. Bacteria and fungi
2. A food source, correct temperature, pH and supply of oxygen if needed
3. To keep the microbe in contact with its substrate
4. Beer, wine, bread, yoghurt, Marmite, mycoprotein (any four)
5. Lactic acid
6. It causes the protein to coagulate (thicken)
7. A fungus
8. Methane
9. A meat substitute produced from a fungus
10. By solving the world's health, food and energy problems

EXAM QUESTIONS - Use the questions to test your progress. Check your answers on page 95.

1. What do we call animals that are hunted by others?

....

2. Put this food chain in the correct order: fox, grass, rabbit.

....

3. With what do food chains always begin?

....

....

4. What is the difference between a producer and a consumer?

....

5. What do we call a diagram showing the numbers of organisms involved in a food chain?

....

6. What is a pyramid of biomass?

....

7. Why are food chains only three or four organisms long?

....

8. List ways in which energy is lost in a food chain.

....

9. Why do the numbers of predators and prey stay relatively stable?

....

10. Give two examples of organisms that have adapted to their environment.

....

11. Why do animals and plants have to adapt to their environment?

....

12. What can happen to the less adapted individuals?

....

13. Give four examples of how humans have reduced the amount of land available.

....

14. What might affect the numbers of a particular animal in a population?

....

....

15. What factors affect the rate of decay when an animal or plant decomposes?

....

....

16. A mammoth was found in frozen soils, why hadn't it decayed?

....

....

17. Approximately how much carbon dioxide is present in the atmosphere?

....

....

18. How do we put carbon dioxide into the air?

....

....

19. What process removes carbon dioxide from the air?

....

....

20. What do plants use the carbon dioxide that they absorb from the air for?

....

21. Give three ways in which acid rain can be reduced.

....

....

22. Give two ways in which the Greenhouse Effect can be reduced.

....

....

23. Whose theory of evolution became accepted?

....

24. What is natural selection?

....

25. What evidence is there for evolution?

....

26. What is biotechnology?

....

27. How is nitrogen returned to the atmosphere?

....

28. Name the types of bacteria in the nitrogen cycle.

....

....

29. How is eutrophication caused?

....

....

30. Outline the process of eutrophication.

....

....

How did you do?

1–7	correct	start again
8–15	correct	getting there
16–22	correct	good work
23–30	correct	excellent

ANSWERS

Humans as Organisms

1. a) Gets smaller; b) Gets larger
2. They carry oxygen around the body.
3. White blood cells.
4. One path from the heart to the lungs and back, and one from the heart to the body and back.
5. At the lungs.
6. At the lungs.
7. Beer, wine and bread.
8. Yeast.
9. Carbohydrates, fats and proteins.
10. Arteries carry blood away from the heart; veins carry blood back into the heart.
11. a) Red blood cells b) White blood cells c) Platelets d) Plasma
12. In the lungs, at the alveoli
13. In the small intestine.
14. To help food move through the gut.
15. The brain, spinal cord and nerves.
16. a) Scurvy; b) Rickets
17. Moist, close contact with the blood, thin membrane, lots of them
18. Aerobic is with oxygen, anaerobic is without oxygen
19. a) Aorta; b) Pulmonary artery; c) Pulmonary vein; d) Vena cava
20. The contracting and relaxing of muscles as food is moved through the gut.
21. Proteases, carbohydrases and lipases.
22. The Biuret test.
23. Ciliated cell; it is found lining the trachea and lungs
24. Ciliary muscles contract, suspensory ligaments slacken, lens gets fatter, object is focused on the retina
25. Diaphragm and intercostal muscles.
26. a) Oxygen; b) Water; c) Energy
27. The liver produces bile, which emulsifies fats (breaks them up into smaller droplets)
28. Sensory, relay and motor neurone.
29. Receptor ➡ sensory neurone ➡ relay neurone ➡ motor neurone ➡ effector ➡ response
30. Glucose ➡ Lactic acid + Energy (a little)

Maintenance of Life 1

1. The seven processes of life: movement, respiration, sensitivity, growth, reproduction, excretion and nutrition
2. Flower, stem, roots, root hairs, leaf
3. To hold the plant upright.
4. Xylem and phloem.
5. Xylem.
6. Phloem: all around the plant particularly to the growing and storage regions.
7. Anchor the plant and take up water and minerals from the soil.

8. Making glucose using sunlight.
9. In the day.
10. All the time.
11. Carbon dioxide + Water ➡ Oxygen + Glucose (using chlorophll to absorb light).
12. Temperature, and the amount of light and carbon dioxide.
13. They contain chlorophyll that absorbs the Sun's energy (light).
14. Long and thin, with plenty of root hairs
15. On the underside of the leaf; to control the opening and closing of the stomata.
16. a) Cell wall; b) Cell membrane; c) Cytoplasm; d) Vacuole; e) Chloroplasts; f) nucleus
17. The movement of molecules from a high concentration to a low concentration. For example, the diffusion of oxygen and carbon dioxide into and out of the stomata.
18. A cell with plenty of water inside its vacuole.
19. A cell that has lost water from its vacuole
20. The flow of water up through the plant and its loss by evaporation from the leaf.
21. Hot, dry and windy conditions.
22. Light, moisture and gravity.
23. A plant hormone found in the shoots and roots of plants.
24. Slow down.
25. The light, because it is leaning towards it and the hormone has gathered on the opposite side.
26. Early ripening of fruit, weedkillers, growing cuttings, seedless fruits
27. Tropisms
28. a) Waxy cuticle; b) Epidermis; c) Palisade cells; d) Spongy layer; e) Leaf vein; f) Stomata; g) Guard cell
29. 1. c; 2. a; 3. b
30. Magnesium and iron.

Maintenance of Life 2

1. a) cell membrane; b) nucleus; c) cytoplasm
2. A plant cell has a cell wall, chloroplasts and a permanent vacuole.
3. They both have a cell membrane, cytoplasm and nucleus.
4. At the lungs (oxygen and carbon dioxide) or the small intestine (digested food).
5. They burst.
6. They have a cell wall.
7. Osmosis
8. Transporting substances against a concentration gradient using energy from respiration.
9. 37°C
10. Brain, nervous system and liver.
11. Emphysema, bronchitis and cancer.
12. Bacteria, fungi and viruses.
13. The body doesn't produce enough insulin/the blood sugar level is too high.

14. By injections of insulin and diet control
15. The pancreas
16. a) Lowers blood sugar levels
    b) Raises blood sugar levels
17. Maintaining a constant internal environment.
18. They dilate, allowing blood to the surface of the skin.
19. Sweating, changes in behaviour.
20. The hypothalamus.
21. A dramatic lowering of body temperature, babies and older people.
22. a) protein coat; b) genetic material (genes)
23. An organism that transports a pathogen from one organism to another e.g. a mosquito.
24. The liver stores excess glucose as glycogen and converts it back to glucose when needed.
25. The product of the breakdown of excess amino acids.
26. In the urine.
27. Filter and clean the blood, reabsorb substances, regulate body water, involved in the formation of urine.
28. Phagocytes and lymphocytes.
29. The ability of the body to remember a disease and produce antibodies to it before symptoms develop.
30. The use of a weak or dead form of a disease given in a vaccine.

Inheritance and Selection

1. 28 days.
2. In the nucleus of body cells.
3. Inheritance and the environment.
4. a) Eye colour, hair colour, blood group, rolling tongues.
   b) Height or weight
5. A shortage of light, water or a low temperature could cause the plant to be shorter than normal, or weaker.
6. 23
7. Exposure to radiation, X-rays, UV light, mutagens
8. Where the best plants or animals are selected by humans to breed from.
9. Asexual reproduction involves one parent and produces clones, sexual reproduction needs two parents and gives rise to variation.
10. The lungs.
11. From two carrier parents.
12. Huntington's chorea.
13. Sickle cell anaemia.
14. Sex cells, sperms and eggs.
15. XX
16. Breed the two strawberries together; take the best offspring and breed together; repeat as necessary.
17. On the chromosomes in the nucleus.
18. It has a stronger affect than the other recessive allele.
19. Phenotype is the physical appearance; genotype is the genes an organism possesses.
20. Clones, in body cells during growth and replacement.
21. In the ovaries and testes, sperms and eggs.
22. 46

23. 46
24. FSH, LH, oestrogen and progesterone.
25. The pituitary gland.
26. FSH
27. 3 : 1 ratio →

	B	b
b	Bb	Bb
B	Bb	Bb

28. Bacteria
29. In the production of insulin.
30. Proteins, protein synthesis.

Environment

1. Prey.
2. Grass, rabbit, fox
3. A plant/producer
4. A producer makes its own food; a consumer eats other organisms.
5. A pyramid of numbers.
6. A pyramid showing the mass of the organisms involved in a food chain.
7. Further up the food chain the energy is lost.
8. Respiration, waste, keeping warm, not all of the organism is eaten
9. The predator–prey cycle.
10. The polar bear and the camel.
11. In order to survive/survival of the fittest.
12. They die out and may become extinct.
13. Quarrying, building, dumping, rubbish, farming
14. Competition, disease, predators, grazing, climate, human activity, amount of food and water
15. Amount of oxygen, temperature and moisture levels
16. Too cold for the bacteria to decay it
17. 0.03%
18. Burning, decomposition, respiration
19. Photosynthesis
20. For photosynthesis and making protein, fats and carbohydrates.
21. Look for alternative energy sources, reduce combustion and the use of cars.
22. Slow down deforestation and the burning of fossil fuels.
23. Charles Darwin's.
24. Nature decides which organisms survive and reproduce.
25. Fossils.
26. The use of plants, animals, bacteria and fungi to produce useful substances and dispose of waste.
27. By denitrifying bacteria that thrive in waterlogged soils, they turn nitrates back into nitrogen.
28. Nitrifying bacteria, nitrogen fixing bacteria, denitrifying bacteria.
29. Leaching of fertilisers into rivers and streams/ excessive rain.
30. Fertilisers are washed into rivers and streams. They cause excessive growth of algae and water plants. The algae die due to competition for light. Decomposers break up the dead plants and use up the oxygen in respiration. Animals die due to oxygen shortage.